Introduction to Agrometeorology and Climate Change

Introduction to Agrometeorology and Climate Change

Alok Kumar Patra
Professor (Agronomy)
All India Co-ordinated Research Project on Integrated Farming Systems
Odisha University of Agriculture & Technology
Bhubaneswar-751 003, Odisha

nipa® books for everyone
PAPERBACKS

New Delhi – 110 034

nipa books for everyone
PAPERBACKS

A Paperback Division of
NEW INDIA PUBLISHING AGENCY
101, Vikas Surya Plaza, CU Block, LSC Market
Pitam Pura, New Delhi 110 034, India
Phone: + 91 (11) 27 34 17 17 Fax: + 91 (11) 27 34 16 16
Email: info@nipabooks.com
Web: www.nipabooks.com

Feedback at feedbacks@nipabooks.com

ISBN: 978-93-87973-61-9

Composed and Designed by NIPA

Preface

Agrometeorology is primarily concerned with the interactions among meteorological, hydrological and pedological factors that influence production systems in agriculture and allied sectors like horticulture, animal husbandry, fishery, forestry, etc.

Since agrometeorology is multidisciplinary in nature it has a well defined approach in theory and applications. Thus, a good understanding of the interrelationships that exist among all the concerned disciplines is required. An agrometeorologist defines all these interactions and correlates physical environments to biological responses and applies the acquired and relevant meteorological skills to help farmers for exploiting weather conditions to improve agricultural production both in quality and quantity.

In this backdrop, agrometeorology has been recommended as a core subject in the curriculum of the agricultural universities. Keeping this in view, an effort has been made to write a textbook on agrometeorology which would be useful to the students as well as to the teachers. This book is primarily based on the syllabus of the course 'Introductory Agrometeorology & Climate Change' meant for under graduate students of agriculture, horticulture and forestry.

In writing this book, the literature on agrometeorology developed by various organisations, agencies and institutes like WMO, FAO, NOAA, BOM, MAFF, IPCC, NASA, IMD, NDMA, CRIDA, ICRISAT, ICAR, AICRPAM, etc. is freely used. I am thankful to the organisations permitting me to quote or cite their literature in the text. I extend my sincere thanks to the authors and editors of various books, journals and periodicals which have been used as reference material in this book. Every care has been taken to cite the bibliographic references. However, any omissions, misrepresentations, incorrect citations or other mistakes that may have occurred are regretted.

I am grateful to my colleagues in the All India Coordinated Research Project on Integrated Farming Systems, Odisha University of Agriculture & Technology, Bhubaneswar for their help in various ways during the preparation of this textbook. I express my gratitude to the New India Publishing Agency, New Delhi for bringing out the book timely and nicely.

I am thankful to my wife Jharashree and daughter Prachurya for their support and encouragement. I also express my indebtedness and gratitude to my beloved parents who are a constant source of inspiration to me throughout my academic journey.

ALOK KUMAR PATRA
OUAT, Bhubaneswar

Contents

Glossary

Abiotic stress: The negative impact of non-living factors on the living organisms in a specific environment.

Absolute humidity: The ratio of the mass of water vapour to the volume of air in which it is present (g/m^3).

Absolute temperature: The temperature of an object on a scale where 0 is taken as absolute zero. Absolute zero is the lowest temperature at which the system is in a state of lowest possible energy and at this point a gas theoretically would cease to exert any pressure. Absolute temperature scales are Kelvin and Rankine.

Absorptivity: The ratio of the amount or radiant energy absorbed to the total energy incident upon the surface.

Acid rain: The rain that has become acidic after contact with certain atmospheric gases, primarily sulphur dioxide, carbon dioxide and nitrogen oxides. This is caused mainly due to air pollution through industrial activity or vehicular exhaust.

Actual evapotranspiration: Rate of evapotranspiration equal or smaller than potential evapotranspiration as affected by the level of available soil moisture, salinity or other causes.

Actual vapour pressure: Pressure exerted by water vapour contained in the air, expressed in millibar (mb) or mm of Hg.

Adaptation: An adjustment by an organism to environmental conditions.

Adaptation to climate change: Initiatives and measures to reduce the vulnerability of natural and human systems against actual or expected climate change effects.

Adiabatic cooling: Cooling of a parcel of gas by expansion, with no heat exchange between the parcel and the surrounding air.

Adiabatic heating: Warming of a parcel of gas by compression, with no heat exchange between the parcel and the surrounding air.

Adiabatic lapse rate: The rate at which temperature changes as air rises or falls. The rate is constant for dry air.

Adiabatic process: A process where a parcel of air cools or warms due to a change in pressure and volume (expansion or compression), with no heat exchange between the parcel and the surrounding air.

Advection: The horizontal transport of moisture, heat or other atmospheric properties by the movement of air.

Advection fog: Fog resulting from the movement of moist air over a colder surface and the consequent cooling of the air to saturation. This type of fog is most common in coastal regions.

Aerosols: Microscopic particles, such as dust, smoke, and bacteria that remain suspended in the air at all times. Aerosols typically only have lifespan of a couple of days as they are continuously washed out of the air by rainfall.

Agricultural drought: Agricultural drought occurs when available soil moisture is inadequate for healthy crop growth and causes extreme stress and wilting.

Agroclimatology: Study of those aspects of climate, which are relevant to the agricultural production system.

Agrometeorological observatory: A place where all the necessary instruments are maintained to observe and record different weather parameters at stipulated time interval.

Agrometeorological station: Station providing simultaneous meteorological and biological information and helping to study and use weather and climate information to enhance or expand agricultural crops and/or to increase crop production.

Agrometeorology: Agrometeorology, abbreviated from agricultural meteorology is an applied science, which deals with the relationship between weather/climatic conditions and agricultural production systems.

Air: The mixture of gases that make-up the earth's atmosphere.

Air mass: An extensive body of air with approximately uniform temperature and moisture characteristics.

Albedo: The ratio of the amount of visible light reflected by a body to the amount incident upon it and expressed in percentage.

Altimeter: An instrument used to determine altitude using atmospheric pressure.

Altitude: The vertical distance from mean sea level to an object aloft.

Altocumulus: A principal cloud type, forming in the middle levels of the troposphere, and appearing as a white and/or grey layer or patch with a waved aspect.

Altostratus: A principal cloud type, forming in the middle levels of the troposphere, and appearing as a grey or bluish sheet or layer of striated, fibrous or uniform appearance, totally or partly covering the sky.

Ambient temperature: The temperature of the surrounding atmospheric air. It is a non-specific phrase used to describe the outside temperature.

Anabatic wind: An uphill wind generated by the heating of a sloping surface. The anabatic wind is also known as upslope wind or valley breeze.

Anemograph: An instrument used to measure both velocity and direction of wind.

Anemometer: An instrument that measures the speed or force of the wind.

Aneroid barometer: A non-liquid instrument used to measure atmospheric pressure.

Anticyclone: An extensive horizontal spiral movement of the atmosphere around and away from a central region of high pressure. The spiral motion is anticlockwise in the southern hemisphere and clockwise in the northern hemisphere.

Anvil: The upper portion of a cumulonimbus cloud that spreads out under the tropopause, often in the shape of a blacksmith's anvil, sometimes for hundreds of kilometres downstream from the parent cloud. It indicates the mature or decaying stage of a thunderstorm.

Aphelion: The time when the earth and sun are at a maximum distance between them.

Arid zone: A zone of very low rainfall with most of the deserts.

Aridity: Refers to excess of evaporation over precipitation.

Aridity index: A numerical indicator of the degree of dryness of the climate at a given location. A number of aridity indices have been proposed; these indicators serve to identify, locate or delimit regions that suffer from a deficit of available water, a condition that can severely affect the effective use of the land for such activities as agriculture or stock-farming.

Atmometer: An instrument used for measurement of evaporation rate.

Atmosphere: The mass of air surrounding the earth and bound to it more or less permanently by the earth's gravitational attraction.

Atmospheric pressure: The force per unit area exerted on a surface by the weight of air above that surface in the atmosphere of earth.

Avalanche: A large mass of rapidly moving snow down a steep mountain slope.

Backing: A counter clockwise shift in the wind direction.

Barogram: The continuous record of atmospheric pressure made by a barograph.

Barograph: Self recording barometer in which continuous trace of atmospheric pressure is made on a barogram.

Barometer: An instrument used to measure atmospheric pressure.

Barotrophic atmosphere: A state of atmosphere where surfaces of constant pressure and constant density coincide.

Beaufort scale: A scale that indicates the wind speed using the effect wind has on certain familiar objects.

Blizzard: A winter storm with strong winds and heavy snow.

Blue watch: In the development of storm watches and warnings, severe thunderstorm warnings are blue or red to indicate the presence of a possible tornado. Severe thunderstorms do not necessarily produce tornadoes, so there is often a decision making process with forecasters to issue a red (tornadic) or blue (non-tornadic) watch. If the watch issued in the forecast has the strong possibility of a tornado, it becomes a red watch.

Broken cloud: An amount of opaque cloud covering the sky of between five and seven oktas.

Calm: The absence of apparent motion in the air.

Cardinal temperature: Below a certain minimum and beyond maximum value of temperatures, plant growth ceases. Between these limits, there is an optimum temperature at which growth proceeds with greatest rapidity. These three points of temperature are the cardinal temperatures for a given plant.

Ceiling: The height of the lowest layer of broken and overcast clouds.

Ceilometer : An instrument for estimating cloud amount and height.

Celsius: A temperature scale in which zero is the freezing point of water and one hundred is the boiling point.

CFC: Acronym for chloro fluoro carbons. Compounds of carbon and halogens released into troposphere where they dissociate to release free chlorine that causes depletion of ozone.

Charge separation: The separation of positive and negative ions into different parts of a cloud. It is a necessary precursor for lightning.

Chilling injury: An injury caused to plants by relatively low temperature above freezing point.

Cirrocumulus: A principal cloud type, forming in the high levels of the troposphere, composed of ice crystals which appear from the ground as very small elements in the form of grains or small ripples.

Cirrostratus: A principal cloud type, forming in the high levels of the troposphere, composed of ice crystals which appear from the ground as a transparent sheet or veil, often creating a halo phenomenon around the sun or moon.

Cirrus: A principal cloud type, forming in the high levels of the troposphere, composed of ice crystals which appear from the ground as white tufts or filaments.

Clear: The appearance of the sky when it is without clouds, or less than 1/10th of the sky is covered by clouds.

Clear air turbulence: A form of turbulence occurring in regions of marked wind shear, particularly at the boundaries of jet streams, but may also be found in strong lee waves. It occurs in the absence of any visual clouds.

Climate: Average meteorological conditions in a certain area over a certain period.

Climate change: A change in the state of the climate that can be identified by changes in the mean and/or the variability of its properties and that persists for an extended period, typically decades or longer.

Climatologist: A climatologist studies long-term weather trends or climatology.

Climatology: The scientific study of long-term atmospheric conditions.

Cloud: An area of the atmosphere containing sufficient concentration of water droplets and/ or ice crystals to be visible.

Cloud base: The bottommost visible portion of a cloud.

Cloud ceiling (height): The altitude of a cloud measured from the earth's surface to the cloud's base up in the atmosphere.

Cloud condensation nuclei: Small particles in the air on which water vapour condenses and forms cloud droplets.

Cloud seeding: A weather modification technique involving the injection of a substance into a cloud for the purpose of influencing the cloud's subsequent development. In common practice, cloud seeding involves the aerial release of silver iodide particles into convective clouds to create thunderstorms.

Cloudy: The state of the sky when 7/10th or more of the sky is covered by clouds.

Cold air advection: The horizontal movement of colder air into a location of warmer air.

Cold front: Transition zone between cold and warm air masses; forms when a cold air mass advances on a warm air mass.

Cold low: A low pressure system with cold air mass from near the surface to all vertical levels.

Condensation: Change of state from vapour to liquid.

Condensation level: The height at which an adiabatically lifted air parcel will become saturated, whereupon condensation occurs.

Condensation nuclei: Tiny airborne particles upon which water vapour condenses.

Conditional instability: Stable unsaturated air that will become unstable if saturated.

Conduction: The transfer of heat between bodies that are in contact, but without any movement of the molecules.

Continental air mass: An extensive body of air, with a more-or-less uniform temperature and moisture profile, which has originated over a large land mass.

Continental climate: The type of climate which is experienced in large landmasses located interior parts of the continent.

Convection: The transfer of heat within a gas or liquid by their movement. In meteorology, it is the vertical transport of heat and moisture, especially by updrafts and downdrafts in an unstable atmosphere. Local breezes, wind and even thunderstorms are a result of convection.

Convective cloud: A cloud that owes its vertical development, and possibly its origination, to convection.

Convective condensation level: The lowest height at which condensation will occur as a result of convection due to surface heating.

Convergence: The condition that exists as a result of a net horizontal inflow of air into a region. Convergent winds at lower levels are associated with upward motion.

Coriolis force (effect): An apparent force on a moving particle that arises solely from the earth's rotation acting as a deflecting force. It acts to the left in the southern hemisphere and to the right in the northern hemisphere. It is greatest at the poles and nonexistent at the equator.

Crop micrometeorology: The branch of meteorology that deals with the interaction of crops and their immediate physical environment.

Cumulonimbus: A cumulus cloud with very deep vertical development extending into the lower stratosphere, distinguished by an anvil at its top consisting of ice crystals.

Cumulus: A principal cloud type having substantial vertical development, forming in the low levels of the troposphere, characterised by flat bases and dome or cauliflower-shaped upper surfaces.

Cyclogenesis: The beginning of cyclone formation.

Cyclone: A region of low pressure relative to the surrounding area. It is a tropical storm originating in the Indian Ocean

Cyclonic rotation: The rotating movement of a tropical storm. In the northern hemisphere, tropical storms rotate counterclockwise. In the southern hemisphere, tropical cyclones rotate clockwise.

Degree day: A measure of the departure of the daily mean temperature from a reference temperature.

Dense fog: A fog in which the visibility is less than 400 m.

Deposition: A process in which a gas transforms into a solid without first becoming a liquid; *i.e.*, water vapour in sub-freezing air changes directly to ice.

Depression: A region of low atmospheric pressure that is usually accompanied by low clouds and precipitation.

Desertification: The development of desert conditions as a result of human activity or climatic changes.

Dew: The water in the form of small liquid drops on grass and other objects at or near the ground formed from condensation of water vapour from the surrounding air.

Dew point (dew point temperature): The temperature at which a parcel of unsaturated air must be cooled in order to reach full saturation at constant pressure and water vapour content. It is the dew point temperature which will cause dew or water droplets to form on surfaces in the mornings.

Dew point front: A dew point front is formed when two warm air masses with differing moisture content meet. Large differences in dew point temperatures create this type of atmospheric boundary.

Diabatic: Processes that involve the addition or removal of heat. For example, air in contact with a cold surface loses heat diabatically by conduction.

Diffuse solar radiation: Sunlight that is scattered downward to the surface.

Direct solar radiation: Sunlight that passes through the atmosphere without absorption or scattering.

Disturbance: A disruption of the atmosphere that usually refers to a low pressure area, cool air and inclement weather.

Diurnal: Pertaining to actions that are completed within twenty four hours, and recur every twenty four hours.

Divergence: Horizontal outflow of air from a particular region.

Doldrums: The area on either side of the equator that experiences low pressures and light shifting winds.

Doppler radar: The weather radar system that employs the apparent shift in frequency of radio waves to perceive air motion and consequently predict tornadoes and precipitation, and measures the speed and the internal characteristics of clouds.

Downburst: A vertical or nearly vertical downward burst of strong winds at ground level usually associated with a thunderstorm.

Downdraft: A descent of cool air associated with convective cloud.

Downslope wind: A wind directed down a slope, caused by greater air density on the slope than at some distance, horizontally from it. The wind is associated with surface cooling of the slope during night.

Drizzle: A slow-falling and uniformly distributed precipitation in the form of tiny water droplets (diameters less than 0.5 mm), usually from stratus or stratocumulus clouds.

Drought: A prolonged period with very little or no rain that leads to a situation when the amount of water available in the soil is not sufficient to meet the demands of potential evapotranspiration.

Drought year: A year in which rainfall of a particular place is short by more than twice the standard deviation.

Dry adiabatic lapse rate: The rate of change in temperature for dry air ascending or descending adiabatically. The rate is approximately 9.8° C/km.

Dry line: The moisture boundary between warm, dry air and warm, moist air. It is a favoured location for thunderstorm development.

Dust devil: A small and rapidly rotating column of wind made visible by the dust, dirt or debris picked up by the wind. It usually occurs in arid or semiarid areas and is most likely to develop on clear, dry, hot afternoons in response to surface heating.

Duststorm: A phenomenon characterised by strong winds and dust-filled air over a large area, and reducing visibility to below one km.

Early warning system: Any series of procedures and devices designed to detect sudden or potential threats to persons, property or the environment at the first sign of danger; especially a system utilising radar technology.

Easterly: An air mass that blows from the east.

Ebb current: The movement of a tidal current away from the coast or down an estuary.

Eddy: A small disturbance in the wind that can produce turbulence.

Effective rainfall: The rain useful for meeting crop-water requirement excluding deep percolation, surface runoff and interception.

***El Nino*:** This is a natural part of the global climate system. It occurs when the central and eastern tropical Pacific Ocean and the atmosphere above it change from their normal state to abnormal warming for several seasons. *El Nino* events occur every three to seven years and are related to shifts in global weather patterns.

Electromagnetic radiation: Energy emitted by virtue of an object's temperature. Radiation does not require a transfer medium and can travel through a vacuum. The energy transfer is accomplished by oscillations in an electric field and a magnetic field.

Elevation: The distance between mean sea level and a point on the earth's surface.

Emissivity: A measure of the efficiency in which a surface emits thermal energy. Emissivity is defined as the fraction of energy being emitted relative to that emitted by a thermally black surface (a blackbody). A blackbody is a material that is a perfect emitter of heat energy and has an emissivity value of 1.

Emittance: The radiant flux density emitted by a unit area of a radiating surface.

ENSO: An acronym for the *El Nino* Southern Oscillation phenomenon. It involves the interaction of tropical Pacific Sea surface temperatures and atmospheric pressures. ENSO refers to the effects of a band of sea surface temperatures which are anomalously warm or cold for long periods of time that develops off the western coast of South America and causes climatic changes across the tropics and subtropics.

Environmental lapse rate: The rate of vertical temperature decrease in the air column. The value is highly variable, depending on local conditions.

Equation of state: The equation relating air pressure to temperature and density.

Equator: The geographic circle at zero degree latitude on the earth's surface. It is an imaginary line around the earth midway between the north and south poles.

Equatorial climate: The summation of weather conditions experienced near the equator which are characterised by high temperature, high humidity, heavy rainfall with little or no seasonal variation.

Equatorial trough: The quasi-continuous belt of low atmospheric pressure located between the subtropical high pressure belts of the northern and southern hemispheres, and moving north and south with the seasons.

Equilibrium level: The height at which a rising parcel of air will become equal in temperature to that of the environment, at which point it is no longer buoyant and thus will cease to rise in the atmosphere without forcing.

Equinoxes: The two days of the year (March 21 and September 23) on which earth's axis is not tilted towards or away from the sun. On the equinoxes every latitude receives 12 hours of sunlight, and sun is overhead at the equator.

Evaporation: The physical process by which a liquid, such as water, is transformed into its gaseous state.

Evaporimeter: Device for measuring the amount of water evaporated into the atmosphere in a given time interval.

Evapotranspiration: The combined processes of evaporation and transpiration.

Exosphere: The outermost layer of the earth's atmosphere (500 to 1000 km above surface); the only part of the atmosphere where an appreciable amount of atmospheric gases escape the earth.

Extended outlook: A basic forecast of general weather conditions three to five days in the future.

Extratropical cyclones: The storms that occur outside of the tropical zones in the mid-latitude regions and higher. These can occur in both the northern hemisphere and the southern hemisphere. The term 'extratropical cyclone' is often used to indicate any regional low pressure system. In the mid-latitudes, temperature differentials on either side of differing air masses can cause large low pressure systems to develop.

Extreme temperature: Highest and lowest temperatures attained during a given time period.

Eye (of a cyclone): An area of clear skies that develops in the centre of a tropical cyclone. It is characterised by light winds and no rainfall.

Eye wall (of a cyclone): An organised band of cumulonimbus clouds that surrounds the eye of a tropical cyclone. It is characterised by high speed winds and most intense rainfall.

Fahrenheit: The standard scale used to measure temperature developed by the German physicist Daniel Gabriel Fahrenheit in 1714; in which the freezing point of water is thirty two degrees and the boiling point is two hundred and twelve degrees.

Fall wind: A strong, cold, downslope wind.

Feeder bands: The lines or bands of cumulonimbus clouds that spiral into and around the centre of a tropical cyclone.

Fetch: The area in which waves are formed by a wind with an approximately constant speed and direction.

Flood: A condition that occurs when water overflows the natural or artificial confines of a stream or river; the water also may accumulate by drainage over low-lying areas.

Foehn: A warm and dry downslope wind descending the lee side of a mountain range.

Fog: A suspension in the air, at or near the earth's surface, of microscopic water droplets, or wet hygroscopic particles, reducing horizontal visibility to less than one km.

Forecast: Something that tells us what the weather is going to be like.

Forecast bust: Whenever a forecast is wrong, or a certain kind of weather is expected to occur but does not, it is called a bust, or a forecast bust.

Freezing level: The altitude in the atmosphere where the temperature drops to zero degree Celsius.

Freezing point: The temperature at which a liquid solidifies under any given set of conditions.

Freezing precipitation: Any form of supercooled precipitation that freezes upon impact with surfaces to form glaze.

Freezing rain or drizzle: The precipitation that falls in liquid form but freezes upon contact with cold objects.

Friction layer (boundary layer, surface layer): The thin layer of the lower atmosphere within which wind direction and speed is subject to frictional effects caused by contact with the earth's surface.

Front: The narrow boundary or transition zone of two air masses with differing properties like temperature and humidity; if cold air advances and replaces warmer air the front is a cold front, and if warm air advances and replaces cooler air the front is a warm front.

Frost: A coating of ice crystals on a surface when the air adjacent to the surface becomes saturated at temperatures below 0° C.

Frozen dew: A coating of ice on a surface that occurs when a layer of dew freezes as temperature drops below 0° C.

Fujita scale: The scale for categorising tornado intensity.

Funnel cloud: A violent, rotating column of air visibly extending towards the earth's surface from the base of a towering cumulus or cumulonimbus cloud. A funnel cloud reaching the ground is called a tornado if over land, and a waterspout if over water.

Gale: A very strong wind with a speed of 63–87 km/hour of sustained surface winds.

Geostationary satellite: An earth-orbiting satellite, placed at an altitude of 35,786 kilometres directly over the equator that revolves around the earth once every twenty four hours in the same direction the earth rotates, thus remaining in the same position relative to the earth's surface.

Geostrophic wind: A wind that blows parallel to straight isobars above the friction layer.

Glacier: A slow-moving mass of ice and snow.

Glaze: A layer or coating of ice that is generally smooth and clear, and forms on exposed objects by the freezing of liquid raindrops.

Global solar radiation: The total solar radiation received on horizontal surface which includes the shortwave radiant energy emitted by the sun and the diffusively scattered solar radiation by the atmospheric components like dust particles, gases, clouds, *etc.*

Gradient wind: A wind that blows parallel to curved isobars above the friction layer due to a constant adjustment between the Coriolis force and the horizontal pressure gradient force.

Gradient: The time rate or spatial rate of change of an atmospheric property.

Grass minimum temperature: The temperature recorded in open air ground on short turf, with the bulb of the thermometer just in contact with the tips of the blades of grass.

Graupel: Ice crystals that have grown by riming to produce a spongy, translucent particle.

Green house effect: The greenhouse effect is a natural process that warms the earth's surface. When the sun's energy reaches the earth's atmosphere, some of it is reflected back to space and the rest is absorbed and re-radiated by greenhouse gases. Greenhouse gases include water vapour, carbon dioxide, methane, nitrous oxide, ozone and some artificial chemicals such as chlorofluorocarbons. The absorbed energy warms the atmosphere and the surface of the earth. This process maintains the earth's temperature at around 33 degrees Celsius warmer than it would otherwise be, allowing life on earth to exist.

Greenwich Mean Time: The local time at zero longitude. The zero longitude is the longitude of Greenwich.

Ground fog: Fog with very little vertical extent (less than 6 m deep).

Growing degree days: A calculation of heat accumulation which is used as an indicator for assessing crop development and maturity. Growing degrees is defined as the mean daily temperature above a certain threshold base temperature accumulated on a daily basis over a period of time. The base temperature is that temperature below which plant growth is zero.

Gust: A sudden and brief increase in wind speed. The duration of a gust is usually less than 20 seconds.

Gust front: The leading edge of cool air rushing down and out from a thunderstorm cloud. It is usually accompanied by a drop in temperature, a wind shift and a pressure jump.

Hadley cell: A large scale wind and pressure pattern found in tropical latitudes of both hemispheres. Air rises above the equator, flows poleward to about 25° latitudes, and flows back to the equator at low levels.

Hail: A precipitation that originates in convective clouds, such as cumulonimbus, in the form of balls or irregular pieces of ice more than 5 mm in diameter.

Hail spike: When moderate to severe hail is associated with a thunderstorm, a long 'spike' of reflectivity usually appears on radar.

Hail storm: A storm, often prolonged and severe, consisting largely of hail or frozen raindrops ranging in diameter from 5 to 10 mm or even more.

Hair hygrometer: An instrument in which the sensing element is a stand or strands of hair and the change of its length is a function of humidity of the air.

Halos: Rings or arcs that seem to encircle the sun or moon and are the result of the refraction of light through the ice crystals that make up cirrus clouds.

Haze: The dust and/or smoke particles suspended in the air, and reducing the visibility to 5000 metres or less.

Heat stroke: A serious heat-related illness that requires immediate medical attention.

Heterosphere: The high atmosphere (about 80 km), where gases are not well mixed but rather are stratified according to molecular weight.

High or high pressure area: An area of high pressure that is usually characterised by clear skies and fair weather carried by sinking air; winds rotate clockwise around and outward from the centre of the high.

High pressure system: An area of pressure maximum with diverging and anti-clockwise winds in the southern hemisphere and clockwise in the northern hemisphere.

High wind: The sustained wind of 65 km/hour or greater and/or gust greater than 93 km/hour.

Homosphere: The lowest 80 km of the atmosphere in which the relative abundance of the permanent gases is constant.

Horse latitudes: The subtropical latitudes between 30 and 35 degrees both north and south. This region, under a ridge of high pressure called the subtropical high, is an area which receives little precipitation and has variable winds mixed with calm.

Humidity: The quantity of water vapour present in the air.

Hurricane: A violent, spiralling tropical storm that forms over the Atlantic Ocean, with winds over 120 km/hour.

Hydrologic cycle: The perennial movement of water in its three phases between the earth, the atmosphere and the seas which includes the change of state and vertical and horizontal transport.

Hydrostatic equilibrium: As the vertical pressure gradient is balanced by the force of gravity, there is no acceleration upward or downward in the atmosphere.

Hygrometer: An instrument used to measure humidity.

Hygroscopic nuclei: Airborne particles having an affinity for water, serving as condensation nuclei.

Hygrothermograph: An instrument that automatically and continuously records both temperature and relative humidity on a single chart.

Ice storm: Liquid rain falling and freezing on contact with cold objects creating ice build-ups of 5 mm or more that can cause severe damage.

Indian standard time: The local mean time for the meridian of longitude 82.5 degree east longitude which passes through Allahabad.

Infrared radiation: Long-wave electromagnetic radiation in higher than visible wavelength segment, between 0.7 and 50 microns.

Insolation: Solar radiation received at the earth's surface. It is a short way to describe incoming solar radiation.

Intergovernmental Panel on Climate Change: The agency of World Meteorological Organisation assigned with assessing and disseminating current knowledge about climate change.

International standard atmosphere: A hypothetical vertical distribution of atmospheric temperature, pressure and density that by international agreement is taken to be representative of the atmosphere for purposes of pressure altimeter calibrations, aircraft performance calculations, ballistic tables, *etc.*

Intertropical convergence zone: The region where southeast and northeast trade winds meet, usually located between 10 degrees north and south of the equator.

Inversion: An increase in temperature with increasing altitude, which is opposite to the usual decrease of temperature with increasing altitude.

Isobar: A line on a weather map linking areas with equal air pressure.

Isohel: A line or curve formed by joining the points of equal sunshine duration over a given period of time.

Isohytes: The line drawn on a weather map joining places which experiences same quantity of rainfall for a given period of time.

Isotach: A line on a chart connecting points of equal wind speed.

Isotherm: A line on a chart connecting points of equal temperature.

Jet stream: A zone of strong winds concentrated in a narrow band in the upper atmosphere, often referred to as the storm track.

Katabatic wind: A wind that carries high density air from a higher elevation down a slope under the force of gravity. Such winds are sometimes also called fall winds.

Kelvin temperature scale: A temperature scale, designed by lord Kelvin of Scotland in 1848, where zero degree K is defined as absolute zero (where all molecular movement stops). Water freezes at 273 K, and boils at 373 K.

Knot: A unit used to measure wind speed equivalent to 1.852 kilometres per hour.

***La Nina*:** A natural part of the global climate system. *La Nina* occurs when the central and eastern tropical Pacific Ocean and the atmosphere above it change from their normal state to sustained cooling for several seasons.

Land breeze: A wind that blows from land out to water along the coastal zone due to temperature differences when the sea is warmer then the adjacent land.

Landfall: When a storm moves from its position over a body of water to land. Tropical cyclones and waterspouts are two kinds of storms that frequently make landfall.

Lapse rate: The rate of change of temperature with height in the atmosphere.

Latent heat: The heat energy that is absorbed when a substance changes from solid to liquid and liquid to gas, and which is released when a gas condenses and a liquid solidifies.

Latitude: A geographic coordinate that specifies the north-south position of a point on the earth's surface. It is measured as the angular distance, subtended at the earth's centre, along a meridian from a point on the earth to the equator. The equator is designated as zero degrees and the poles as 90 degrees.

Lee (leeside or leeward): The side of an obstacle (usually of a mountain) that is protected from the prevailing wind.

Lenticular cloud: A more or less isolated cloud, downwind of a barrier, resembling a smooth lense with sharp outlines. They mostly occur in mountain waves, and thus indicate possible severe turbulence and icing.

Level of free convection: The height at which a parcel of saturated air becomes warmer than the surrounding air and thus begins to rise freely until it reaches its equilibrium level.

Lifting condensation level: The height at which a parcel of moist air becomes saturated when it is cooled by adiabatic lifting.

Lightning: A visible electrical discharge produced by a thunderstorm. It can occur between cloud and ground, between clouds, within a single cloud, or between a cloud and surrounding air.

Long range forecast: A weather prediction extending beyond 7 days.

Longitude: A geographic coordinate that specifies the east-west position of a point on the earth's surface. It is measured as an angle in reference to the prime meridian, which is designated as zero degrees longitude.

Low (low pressure system): Area with lower atmospheric pressure than its surrounding areas. This makes air from surrounding areas to flow into the low and may lead to cloudiness and precipitation.

Low latitudes: The latitude belt between the equator and 30 degrees north and south of the equator.

Low level jet: A transient strong wind concentrated in relatively narrow bands near the earth's surface.

Maritime air mass: An air mass that has originated over an extensive water surface and usually has a uniform temperature and moisture profile.

Maximum sustained surface wind speeds: A sustained wind is one that is relatively long lasting. In most parts of the world, a 10 minute time period is used to determine sustained winds.

Maximum thermometer: A thermometer used for measuring the highest temperature attained during a day.

Measurable precipitation: Precipitation whose liquid water equivalent is 0.25 mm or greater. For frozen precipitation, accumulations are melted first and then the resulting water is measured.

Medium range weather forecast: Weather forecasts for agriculture, predictions for 3 to 7 days in advance.

Mesopause: The uppermost boundary of the mesosphere with the lowest temperature in the atmosphere.

Mesoscale: A scale of meteorological phenomena typically having horizontal extents of several tens of kilometres.

Mesosphere: Region of the atmosphere from about 50 km to 80 km, characterised by decreasing temperature.

Meteorology: The study of phenomena of the atmosphere and all the processes that take place in the atmosphere and their relationships with processes at the surface of the earth.

Microburst: A severe downburst of wind, usually from a thunderstorm, over an area of less than four kilometres in diameter. The term downburst is used to refer to larger diameters.

Microclimate: A local climate that differs from the main climate around it.

Micrometeorology: The study of local and small-scale atmospheric phenomena usually confined to the physical and dynamic occurrences within a shallow stratum of air adjacent to the ground.

Microscale: The smallest scale of meteorological phenomena.

Millibar: The scientific unit used in measuring and reporting atmospheric pressure, abbreviated as mb.

Minimum thermometer: A thermometer used for measuring the lowest temperature attained during a day.

Mist: A suspension in the air, at or near the earth's surface, of microscopic water droplets or wet hygroscopic particles which reduce the horizontal visibility to less than five km but not less than one km.

Mixing ratio: A measure of atmospheric moisture; the mass of water vapour per unit mass of dry air, usually expressed in g/kg.

Monsoon: A seasonal wind that brings rain to many places of the world.

Monsoon depressions: Areas of low pressure superimposed in the southeasterly airflow out of the Bay of Bengal.

Mountain breeze: A breeze that flows down a hill at night.

Net solar radiation: The difference between the radiation received from the sun and the solar radiation reflected back upwards.

Nimbostratus: Low level thick dark cloud with more or less continuously falling rain, snow or sleet.

Numerical weather prediction (NWP): The process of using mathematical equations to estimate the atmosphere's state at some point and time in the future.

Occluded front: A front formed when a cold front overtakes a warm front.

Occlusion: It occurs as a result of the collision of two frontal boundaries in the atmosphere. An occlusion is the process of formation of an occluded front.

Ocean current: The horizontal movement of surface waters caused by prevailing winds.

Okta: A unit of measure to describe how much of the sky is covered in cloud. One okta is equal to one eighth of the total exposed hemisphere of the sky. The sky conditions are estimated by dividing the sky into eight slices and visually observing how many of those slices' area is covered in cloud.

Orbit: The path of the earth around the sun, or of satellites around the earth.

Orographic lifting: The rising of flow of air over physical barriers such as mountains.

Overcast: In order to classify the sky as overcast, 90 to 100% of the sky is considered to be covered by clouds. The penetration of sunlight is noticeably less.

Ozone layer: A layer of air in the atmosphere that contains a relatively high concentration of the gas ozone. It absorbs most of the ultraviolet radiation from sun which are harmful to humans as well as other life forms on earth.

Parcel: A theoretical small and self-contained volume of air responding to meteorological processes as a single entity.

Perihelion: Earth's closest approach to the sun (on about January 3).

Permafrost: A soil layer below the surface of tundra regions that remains frozen permanently.

Photoperiod: The relative length of time that a plant is exposed to light especially as it affects the life cycle and physiological processes of the plant.

Photoperiodism: The effect of day length on flowering, dormancy, vegetative development, leaf fall and other aspects of plant behaviour.

Photosynthetically active radiation: The radiation that exists in the waveband between 0.4 to 0.7 microns.

Physiological drought: Non-availability of water to the plant due to unfavourable physiological conditions such as water logging, soil salinity, low temperature, *etc.* even though there is sufficient water in the soil.

Pilot balloon: A balloon of relatively small size used to collect the data on winds.

Polar air mass: An air mass that forms over a high latitude region. Continental polar air is formed over cold land surfaces and is typically very stable with low moisture.

Polar easterlies: Low-level winds originating in the polar highs.

Polar front: Transition zone between cold polar air and warmer air of the mid-latitudes.

Polar front theory: The theory postulated in the early part of the twentieth century describing the formation, development and dissipation of mid-latitude cyclones.

Polar highs: Low-level anticyclones of the Arctic and Antarctic.

Polar jet stream: A jet stream found in the upper troposphere above the polar front, a result of the strong temperature contrast across the front.

Polar-orbiting satellite: A satellite whose orbit passes over both of the earth's poles.

Potential evapotranspiration: The amount of evapotranspiration that may occur if plenty of water is present in soil. It includes both possible maximum evaporation from the soil and transpiration from the vegetation of a specified region in a given time interval when water is not a limiting factor.

Precipitation: Any product of the condensation of atmospheric water vapour that falls from the sky and reaches ground under gravity.

Pre-frontal squall line: A line of thunderstorms that may develop ahead of an advancing cold front, and having an orientation more or less parallel to the cold front.

Pre-frontal trough: An elongated area of relatively low pressure that may develop ahead of an advancing cold front.

Pressure: A measure of the weight of the air that is usually measured with a barometer.

Pressure altimeter: An aneroid barometer calibrated to indicate altitude by measuring atmospheric pressure and using the standard atmosphere pressure/height relations.

Pressure centre: A region of high or low pressure.

Pressure gradient: The change in atmospheric pressure per unit horizontal distance in the direction in which pressure changes most rapidly.

Pressure gradient force: A force that arises from spatial variation in pressure. It causes air to blow from an area of high pressure towards an area of low pressure.

Prevailing wind: A wind that blows from one direction more frequently than any other during a given period.

Prognostic chart: A weather forecast chart.

Psychrometer: An instrument for measuring the atmospheric moisture.

Pyranometer: An instrument for measuring the total incoming radiation.

Pyregeometer: An instrument for measuring the effective terrestrial radiation.

Pyroheliometer: An instrument for measuring the intensity of direct solar radiation at normal incidence.

Quasi-stationary front: A front which is nearly stationary or moving very slowly.

Radar: Acronym for radio detection and ranging. An electronic instrument used to detect distant objects and measure their range by detecting scattered or reflected radio energy.

Radiation: Energy that comes from a source and travels through some material or through space. Light, heat and sound are types of radiation.

Radiation fog: Fog that forms when radiational cooling at the earth's surface lowers the temperature of the air near the ground, to or below its dew-point.

Radiational cooling: The cooling of the earth's surface and the adjacent air which occurs at night when the earth's surface suffers a net loss of heat due to outgoing radiation being greater than incoming radiation.

Radiometer: An instrument used to measure the intensity of solar radiation.

Radiosonde: A balloon carrying instruments for measuring conditions in the upper atmosphere.

Rain: Precipitation in the form of liquid water droplets greater than 0.5 mm diameter.

Rain gauge: Instrument used to measure the depth of water from precipitation.

Rainshadow: An area on the leeward side of a mountain barrier having comparatively low precipitation.

Rainy day: A twenty four hour period, usually starting from 0730 IST during which a minimum amount of 0.2 mm of precipitation.

Rawinsonde: A balloon that is tracked by radar to measure wind speeds and wind directions in the atmosphere.

Rayleigh scattering: The scattering of radiation by agents substantially smaller than the radiation's wavelength. In the case of the atmosphere, this applies to the scattering of visible radiation by air molecules.

Refraction: The bending of any wave as it passes from one medium to another, in which the speed of travel is different.

Relative humidity: A ratio of the actual amount of water vapour (moisture) present in the air 'relative to' the amount of water vapour that would be present if air at that same temperature and pressure was saturated. Relative humidity is expressed as a percentage.

Remote sensing: Acquisition of physical data of an object without touch or contact.

Ridge: An elongated high pressure area in the atmosphere that lies between two low pressure areas.

Rime: Ice formed by the rapid freezing of supercooled water droplets when they contact an exposed object, forming a white and opaque granular deposit of ice.

Roll cloud: A low-level, horizontal, tube-shaped cloud usually associated with a thunderstorm gust front, where the roll cloud is completely detached from the base of the cumulonimbus cloud.

Saffir-Simpson scale: The scale measures the intensity of a hurricane, namely the level of damage it is expected to cause to structures on a scale of 1 to 5 (5 being the most intense).

Sandstorm: The strong wind carrying sand through the air and reducing visibility to less than 1 km.

Satellite: Device that circles high above the earth and can be used for monitoring the weather.

Saturate: To add something to the point where no more can be absorbed, dissolved, or retained. In meteorology, it is used when discussing the amount of water vapour in a volume of air.

Saturated adiabatic lapse rate: The rate of change in temperature of a saturated air parcel as it adiabatically ascends or descends through the atmosphere. The value ranges from about 4 to 10° C/km, depending mainly on temperature.

Sea breeze: A coastal breeze that blows from sea to land during daytime in summer that is caused by warm air rising over land and being replaced by cooler air from over the sea.

Sea surface temperature: The average water temperature between the ocean's surface and a depth of 20 m below it.

Sensible heat: The energy contained in air that can be sensed *via* its temperature.

Severe thunderstorm: A thunderstorm that produces very strong winds, large hail, or tornadoes.

Shower: Precipitation from a convective cloud that is characterised by its sudden beginning and ending, changes in intensity, and rapid changes in the appearance of the sky. It occurs in the form of rain, snow, or hail.

Sleet: Precipitation in the form of very small ice pellets, resulting when drizzles freeze before reaching the surface.

Smog: A mixture of smoke and fog; a natural fog contaminated by industrial pollutants.

Solar constant: The amount of radiation reaching the top of the atmosphere when earth is at its average distance from the sun. This is not a pure constant, but rises and falls with changes in solar emission. Its value is about 1368 W/m^2.

Solar wind: A continuous stream of particles (mostly protons and electrons) emitted by the sun, travelling about one-third to one-half the speed of light.

Solstices: The two times each year (June 21 and December 22) that mark the northern and southern limits of the latitude of overhead sun. On the June solstice, northern hemisphere has its longest day of the year and on December solstice southern hemisphere has its longest day.

Southern oscillation: The reversal of surface pressure pattern over the tropical Pacific associated with *El Nino* events.

Southern oscillation index: An indication of the development and intensity of *El Nino* or *La Nina* events in the Pacific Ocean. The SOI is calculated using the pressure differences between Tahiti and Darwin.

Squall: A sudden onset of strong winds with speeds increasing by at least 30 km/hour and sustained at 40 km/hour or more for at least one minute. The intensity and duration is longer than that of a gust.

Stable air: Air that resists uplift. When displaced vertically, it returns to its initial position.

Standard atmosphere: A mathematical model of the atmosphere which is standardised so that predictable calculations can be made.

Standing wave: An atmospheric wave that is stationary with respect to the earth's surface.

Stationary front: The boundary between cool and warm air in which neither air mass is advancing.

Steam Fog: Fog formed when cold air moves over a warmer water surface.

Stefan-Boltzmann law: The law for blackbody emission stating that the total energy emitted over all wavelengths is proportional to the fourth power of absolute temperature.

Stevenson screen: A white box with ventilated sides that is used to house weather instruments and protect them from direct sunlight

Storm surge: A potentially damaging influx of coastal waters brought about by high winds and low pressures associated with hurricanes.

Storm track: The path that a low pressure area follows.

Stratiform: Clouds that exhibit extensive horizontal development.

Stratocumulus: A principal cloud type, forming in the low levels of the troposphere and existing in a relatively flat layer but having individual elements, from which drizzle can fall.

Stratopause: The boundary zone between the stratosphere and the mesosphere. In the stratosphere the temperature increases with height, with the stratopause being the point of maximum temperature, prior to a decrease in temperature in the mesosphere.

Stratosphere: A layer of the atmosphere between about 16 and 50 km characterised by generally increasing temperature with increasing altitude and an absence of convective clouds and associated turbulence.

Stratus: A principal cloud type, forming in the low levels of the troposphere and normally existing as a flat layer that does not exhibit individual elements.

Sublimation: Change from solid phase into a gaseous phase without passing through the liquid phase; *i.e.*, the process of an ice changing directly into water vapour.

Subsidence: A descending motion of air in the atmosphere, usually with the implication that it extends over a broad area.

Subtropical jet: A band of relatively strong winds found between 20 and 40 degree latitude in the middle and upper troposphere. Jet streams form near boundaries of adjacent air masses with significant differences in temperature.

Supercooled water: Water in liquid state even though it has been cooled to a temperature below its normal freezing point.

Supersaturation: When the relative humidity is greater than 100% *i.e.*, the atmosphere is more than saturated with water vapour. It requires a very clean atmosphere, where condensation nuclei are lacking.

Synoptic chart: Any chart that depicts meteorological or atmospheric conditions over a large area at a given time.

Synoptic scale: The scale of meteorological phenomena having areas on the order of hundreds or thousands of square kilometres.

Temperate zone: The latitudinal zones on earth's surface which lie between 23.5° and 66.5° in both the hemispheres.

Temperature: A measure of how hot or cold something is.

Temperature gradient: A physical quantity that describes in which direction and at what rate the temperature changes the most rapidly around a particular location. The temperature gradient is a dimensional quantity expressed in units of degrees (on a particular temperature scale) per unit length. A strong temperature gradient implies that temperature changes rapidly over a short distance.

Temperature inversion: Condition in which temperature increases with increasing altitude in the atmosphere in contrast to the normal decrease with altitude. When temperature inversion occurs, cold air underlies warmer air at higher altitudes.

Thermal low: Low-pressure cell produced by heating of the surface.

Thermal radiation: The infrared radiation emitted by the earth and its atmosphere including material at terrestrial surface.

Thermistor: An object whose electrical resistance changes with temperature, thus allowing temperature to be determined by measuring changes in electrical current.

Thermodynamic diagram: A diagram showing the relationship among pressure, temperature, density and water vapour content, such that characteristics of air parcels can be determined as they ascend and descend.

Thermograph: A device used to give continuous record as a geographical chart of temperature with time.

Thermohaline circulation: A movement of surface waters in the oceans due to variations in temperature and salt content.

Thermohygrograph: A device used for measurement of both atmospheric temperature and humidity.

Thermometer: An instrument for measuring the temperature of something, for example the air.

Thermoperiodism: Effects of alternations of temperature between the day and night periods upon the reactions of plants.

Thermosphere: the layer of the atmosphere located between the mesosphere and outer space. It is a region of increasing temperature with height, and includes all of the exosphere and most of the ionosphere.

Thunder: Sound produced when lightning discharges heat to the surrounding air, causing pressure waves to emanate outward.

Thunderstorm: A cumulonimbus cloud characterised by thunder and lightning and associated gusty surface winds, hail, rain, turbulence, icing and, under the most severe conditions, microbursts and/or tornadoes.

Tornado: A rotating column of air extending between a cumulonimbus cloud and the ground. It is the most destructive of all storm-scale atmospheric phenomena.

Tornadogenesis: The processes and conditions necessary for the formation of tornadoes.

Torrential rain: A rapid and violent rainfall that may cause economic losses.

Trace: Amounts of rain or other forms of precipitation that is less than 0.25 mm.

Trade winds: are easterly winds that constantly blow across the tropics. These winds are generated by the difference in pressure between the subtropical high located at 30° N and 30° S latitude and the equatorial low at 0°. There are two trade belts; one blowing from the northeast in the northern hemisphere, and another blowing from the southeast in the southern hemisphere. Trade winds from both hemispheres meet near the equator.

Transpiration: Transfer of water to the atmosphere by vegetation, mostly by water evaporating and escaping plant tissues through leaf pores.

Tropic of Cancer: A line of latitude at 23.5° N, the northern limit of solar declination.

Tropic of Capricorn: A line of latitude at 23.5° S, the southern limit of solar declination.

Tropical air mass: An air mass that forms in the tropics or subtropics. Maritime tropical air is produced over oceans and is warm and humid, while continental tropical air is formed over arid land and is very hot and dry.

Tropical cyclone: A non-frontal low pressure system that develops over tropical waters with a wind of more than 60 km/hour.

Tropical depression: A non-frontal low pressure system that develops over tropical waters with wind speeds of 60 km/hour or less.

Tropical disturbance: An area of organised convection, originating in the tropics or occasionally the subtropics, that maintains its identity for 24 hours or more, but has no closed wind circulation. It is often the first developmental stage of a tropical cyclone.

Tropical region: The area between the Tropics of Cancer (23.5° N latitude) and Tropics of Capricorn (23.5° S latitude).

Tropical storm: A storm that originates in tropical regions and has wind speeds between 60 and 120 km/hour.

Tropopause: The boundary zone or transition layer between the troposphere and the stratosphere.

Troposphere: The lowest region of the atmosphere in which all the earth's weather takes place.

Trough: An elongated relatively low pressure area between two high pressure areas.

Turbulence: An irregular motion of air caused by eddies in contrast to a smooth laminar flow.

Typhoon: The name for a tropical storm originating in the Pacific ocean, usually the China Sea.

Ultra-violet radiation: The radiation found in the sun's rays beyond the violet part of the visible spectrum, having wavelength range from 0.005 to 0.4 micron. It can be harmful to both plant and animal life

Unstable atmosphere (air): An atmosphere in which air parcels rise buoyantly due to the rising air parcel being less dense (warmer) than the surrounding air.

Updraft: A current of air with marked vertical motion produced by the process of convection. If the air has sufficient moisture, it will condense and form a cumulus cloud. If updrafts continue to build upward, they will grow a cumulonimbus (thunderstorm) cloud.

Valley breeze: A low level movement of air in an upslope direction, developing during daylight hours as a result of solar heating.

Vapour pressure: The partial atmospheric pressure exerted by water vapour. It is a measure of atmospheric moisture.

Vapour pressure deficit: The difference between the actual vapour pressure in the crop canopy and the vapour pressure at saturation.

Veering wind: The wind that changes the direction in a clockwise shift.

Virga: Rain falling from a cloud that evaporates before reaching the ground.

Visibility: A measure of the opacity of the atmosphere. It is the greatest distance one can see prominent objects with unaided normal eyesight.

Vorticity: The turning of an object such as an air parcel, usually with respect to the vertical direction. It is associated with areas of divergence and convergence.

Walker circulation: A deep east-west circulation induced by the contrast between the warm waters of the western Pacific and the cooler waters of the eastern Pacific.

Warm air advection: The horizontal movement of warmer air into a location.

Warm front: Transition zone between two air masses of different temperature, produced when warm air advances on and overruns cold air.

Watch: Information to the public that conditions are favourable for the occurrence of severe weather conditions within twelve to thirty-six hours of the given announcement.

Water vapour: Water in its gaseous state that comprises one of the most important constituents of the atmosphere.

Waterspout: A rather weak whirlwind (narrow rotating column of air), forming over a water surface. Rising and condensing air makes the waterspout visible.

Wavelength: The distance between two successive wave crests separated by a trough.

Weather: Day-to-day conditions of the atmosphere.

Weather balloon: A weather instrument used to measure conditions such as air pressure, temperature, humidity, and wind data above earth's surface. It consists of a measuring device, or radiosonde, which measures weather aloft and transmits this data back to the surface, and a high-altitude balloon made of highly flexible latex and filled with hydrogen gas, responsible for lifting the radiosonde aloft.

Weather synopsis: A description of weather patterns affecting a large area.

Westerlies: The Westerlies or anti-trade winds are prevailing winds from the west towards the east in the middle latitudes between 30 and 60 degrees latitude of both the hemispheres. They originate from the high-pressure areas in the horse latitudes and tend towards the poles.

Wet bulb temperature: The temperature recorded on a thermometer whose bulb is surrounded by a wet muslin cloth, thus lowering the temperature by loss of latent heat through evaporation.

Wet bulb thermometer: One of the two thermometers of a psychrometer whose bulb is wet with thin film of distilled water and indicates wet bulb temperature.

Wet year: If the rainfall exceeds twice the normal deviation at a particular place, the year is said to be wet year.

Whirling psychrometer: Psychrometer attached to a string or small chain which an observer can rotate in order to ensure good ventilation.

Whirly wind: A small-scale, rapidly rotating column of wind, formed thermally and thus most likely to develop on clear, dry and hot afternoons. Often called a dust devil when made visible by the dust, dirt or debris it picks up.

Wien's law: Law for blackbody emission that states that the wavelength of maximum emission is inversely proportional to absolute temperature.

Wind: A moving mass of air. The movement of the air causes the weather.

Wind chill temperature index: An index of apparent temperature used for cold conditions that incorporates air temperature and wind speed.

Wind direction: The direction from which the wind is blowing.

Wind profiler: A type of Doppler radar unit that provides information about vertical changes in wind speed and direction.

Wind shear: Large change in wind direction and/or speed over a short vertical or horizontal distance.

Wind vane: An instrument for measuring or indicating wind direction, often a pivoting arrow that continually points into the wind.

Windward: This is the direction from which wind is blowing.

WMO: Stands for the World Meteorological Organisation, the agency of the United Nations charged with collecting and disseminating meteorological data.

X-band: A frequency band of microwave radiation in which radars operate.

Yellow wind: A strong, cold, dry west wind of eastern Asia that blows across the plains during winter and carries a yellow dust from the desert.

Zenith angle: The angle between the sun and the vertical direction. When the sun is overhead, the zenith angle is zero.

Zonal wind: The wind flowing east-west, parallel to a line of latitude.

Chapter 1

Introduction

Meteorology is derived from the Greek word *meteoro* which means above the earth's surface; *i.e.*, the atmosphere. The state of the atmosphere experienced at a given time, in any location on the surface of the earth is known as weather. Thus meteorology is the scientific study of the atmosphere or the weather. It is a branch of physics dealing with atmosphere and it is often quoted as the 'physics of the lower atmosphere'. The main focus of meteorology is to understand the weather processes and forecast the weather. Solar radiation, the air mass and moisture are the three key ingredients that make up weather. The exchange of heat, moisture and momentum between the earth's surface and the atmosphere determines the characteristic of weather. The observable weather events are known as meteorological phenomena and are defined by a set of variables called weather elements. Atmospheric temperature, atmospheric pressure, humidity, precipitation, cloudiness, wind and visibility, *etc.* are examples of such weather elements. The weather elements are not separate entities. They are closely related with each other and are highly variable in nature. Meteorology explains and analyses the changes of these weather elements.

METEOROLOGY

The major subdivisions of meteorology are dynamic meteorology (or atmospheric dynamics) and physical meteorology (or atmospheric physics). The dynamic meteorology is concerned with the analysis and interpretation of the three-dimensional, time-varying, macro scale motions and energy transformations in the atmosphere. The science of dynamic meteorology is meant to develop models based on the hydrodynamic and thermodynamic equations. The physical meteorology is concerned with the processes that alter the physical properties and the chemical composition of air parcels as they move through the atmosphere.

Depending on the temporal and spatial scope of interest, meteorology can be divided into distinct areas of emphasis such as micro, meso, synoptic and planetary scale meteorology. These scales are related to the size of the atmospheric motion and the spatial size of each of them corresponds to appropriate timescale. Micro-scale meteorology pertains to short-lived atmospheric phenomena which last for minutes or hours rather than days and covers a distance of about one km or less. It controls the most important mixing and dilution processes in the atmosphere. Important topics in micro-scale meteorology include heat transfer and gaseous exchange between soil, vegetation and/or surface water and the atmosphere caused by near-ground turbulence. Most of the micro-scale phenomena occur within the planetary boundary layer, where the frictional force of ground topography can influence the air motion. Individual cloud systems at higher levels also come under the purview of micro-scale meteorology. Meso-scale meteorology ranges in horizontal dimensions from 5 to 500 kilometres and typically possesses lifetimes of a day or less. Vertical velocity often equals or exceeds horizontal velocities in meso-scale meteorological systems. Examples of meso-scale weather systems are tornadoes, thunderstorms, cloud bursts, land and sea breeze, *etc*. Synoptic (or cyclonic) scale meteorology has a horizontal length scale of the order of 500-3000 km and lasts for several days. Synoptic scale meteorology is best studied through the air mass concept. Synoptic scale weather systems are the major features shown on the weather maps such as mean sea level surface charts and the daily forecast charts. They are characterised by mainly horizontal motion. Tropical and extra tropical cyclones are the best examples of synoptic scale systems. Planetary (or global) scale meteorology studies weather patterns related to the transport of heat from the tropics to the poles and other very large scale oscillations. The atmosphere acts like a gigantic heat engine and the temperature difference between the poles and the equator provides the energy supply needed to drive planetary atmosphere and ocean circulation. Planetary scale systems last from several days to weeks, and extend over several thousand kilometres.

CLIMATE AND WEATHER

Climatology is derived from the Greek word *klima* which means slope of the earth. Climatology is the scientific study of climate, which is usually defined as the average weather. It seeks to describe and explain the nature of the climate and how it differs from place to place. Climate can be defined as the state of the atmosphere expressed over a longer period of time (generally 30 years) and includes a statistical description in terms of the mean and variability of relevant weather elements. It really represents a kind of frequency distribution of weather types in a given region and over a period of time. Until mid-twentieth century, climate was considered to be sufficiently constant and the average weather of 30 years at any location usually represented the climate of that location. However, the variation of climate averaged over time is now widely acknowledged.

The distinction between weather and climate can be appreciated with the popular quotations '*Climate lasts all the time, weather only for few days*' (Mark Twain) or '*Climate is what you expect, weather is what you get*' (Robert Heinlein). Weather refers to the state of atmosphere at any given time denoting the short-term variations of atmosphere in terms of temperature, pressure, wind, moisture, cloudiness, precipitation, *etc*. It is highly variable, constantly changing, sometimes from hour to hour and at other times from day to day. These weather elements are closely related with each other. Climate on the other hand, is the sum total of the varieties of weather conditions of a place or an area. It may be defined as the sum of all statistical weather information of a particular area during a specified interval of time, usually several decades. The World Meteorological Organisation has suggested a standard period of 31 years for calculating the climatic averages of different weather elements.

Meteorology and climatology are two closely related fields of study. Both of these study the elements of atmosphere and the causes of atmospheric behaviour. Meteorology focuses on short term weather systems lasting up to a few weeks. On the other hand climatology studies the frequency and trends of those systems. It studies the periodicity of weather events over years to millennia, as well as changes in long-term average weather patterns, in relation to atmospheric conditions. Meteorology draws its fundamental principles from physics and applies them to the behaviour of the atmosphere, which follows certain basic rules. Changes in the pattern of heat, moisture and motion in three dimensional spaces can be described and accounted for by the laws of physics. Meteorology treats weather processes as a set of physical changes and can be considered as a branch of physics, whereas climatology is concerned with the weather process as the expression of regional and seasonal influences and often is considered as a part of geography. Due to the present concern about global warming related climate change, the areas of research in both meteorology and climatology are rapidly growing and many overlapping areas of multidisciplinary nature are being developed. Thus, the distinction between the two subjects is getting blurred and it is advantageous to treat meteorology and climatology together.

Climate can be considered under micro, meso and macro levels. Climatic conditions in a relatively small area, within a few feet above and below the earth's surface and within canopies of vegetation is called microclimate. Microclimates are affected by not only the weather elements such as temperature, humidity, wind and turbulence, dew, frost, heat balance, evaporation, but also by the nature of soil and vegetation, the local topography, elevation and season. The climate of small areas of the earth's surface which may not be representative of the general climate of the district is known as meso-climate. It is controlled by local factors. The large-scale climate of a large area or country, as distinguished from meso-climate and microclimate is known as macro climate. Any portion of the earth's surface over which the broad climatic characteristics are similar is called a climatic region. It can be further grouped into climate types and climate groups.

AGROMETEOROLOGY

Agrometeorology, abbreviated from agricultural meteorology is a branch of applied meteorology, which investigates the physical conditions of the environment for growing plants, animals or other organisms. It is an applied science, which deals with the relationship between weather/climatic conditions and agricultural production or it is a science concerned with the application of meteorology to the measurement and analysis of the physical environment in agricultural systems.

Agrometeorology is defined as the science investigating the meteorological, climatological, and hydrological conditions that are significant to agriculture owing to their interaction with the objects and processes of agriculture production (Molga, 1962). Agrometeorology puts the science of meteorology to the service of agriculture in its various forms and facets to help with the sensible use of land, to accelerate the production of food, and to avoid the irreversible abuse of land resources (Smith, 1970). Agrometeorology is concerned with meteorological, hydrological and pedological factors affecting agricultural production and also with agriculture's interaction with the environment. Its objectives are to elucidate such effects, and then to assist farmers to apply this supportive knowledge and information in agrometeorological practices and services.

Agrometeorology is an interdisciplinary science in which the main scientific disciplines involved are atmosphere sciences and soil sciences, which are concerned with the physical environment, and plant sciences and animal sciences (including their pathology, entomology, and parasitology, *etc.*), which deal with the contents of the biosphere (Mavi and Tupper, 2004). Though interdisciplinary in nature, agrometeorology is a well-defined science. Its subject matter links together the physical environment and biological responses under natural conditions. An agrometeorologist applies every relevant meteorological skill to help farmers make the most efficient use of their physical environment in order to improve agricultural production both in quality and quantity and to maintain the sustainability of their land and resources (Bourke, 1968). Using a four-stage approach, an agrometeorologist first formulates an accurate description of the physical environment and biological responses. At the second stage, he or she interprets biological responses in terms of the physical environment. Third, he or she makes agrometeorological forecasts. The final goal is to develop agrometeorological services, strategies, and support systems for on-farm strategic and tactical decisions and to implement them in collaboration with specialists in agriculture, livestock, and forestry (Mavi and Tupper, 2004).

SCOPE AND IMPORTANCE OF AGROMETEOROLOGY

Among all the human activities, agriculture is possibly the most dependent on the weather. Weather and climate are natural resources and are considered as basic inputs in agricultural planning. Climate influences the distribution of crops over different regions of the world, while weather influences crop production and productivity. All the farm operations from land preparation to harvesting, threshing, storage and transport of farm produce are affected by weather. The field of interest of an agrometeorologist extends from the depth down to which tree roots penetrate to the highest levels in the atmosphere through which the transport of seeds, spores, pollen, and insects may take place. Agrometeorology is concerned with the interactions among meteorological, hydrological and pedological factors on the one hand and agriculture, in the broader sense including horticulture, animal husbandry, fishery, and forestry, on the other. Thus, the role of an agrometeorologist is to find out and define these interactions, and then apply this acquired knowledge in the operational field of agriculture and allied sectors to increase the productivity. All the agricultural activities are affected directly or indirectly by weather and climate. Agrometeorology has therefore, the greatest scope on agriculture and other allied enterprises as summerised below.

Characterisation of agroclimate: The growth, development and yield of a crop depend on the climatic parameters such as solar radiation, air temperature, precipitation, wind, humidity, *etc.* Agrometeorology considers and assesses the suitability of these parameters in a given region for maximum crop production and economical benefits. The meteorological data supplemented with soil and topographic information help to develop a suitable land use planning.

Crop planning for stability in production: To get stabilised yields even under weather adversity, suitable cropping patterns/ contingent cropping planning can be adopted by considering water requirements of crops, effective rainfall and available soil moisture. This may reduce the risk of crop failure due to abiotic stresses.

Crop management: Farm operations such as sowing, fertiliser application, plant protection, irrigation scheduling, harvesting, drying and processing, *etc.* are carried out more effectively and scientifically if the prevailing or forecasted weather conditions are available to the farmers. Weeding, harrowing, mulching, *etc.* are undertaken during dry spells forecasted. Similarly, fertiliser application is advisable when rainfall is not heavy and soil moisture is between 30 and 80%. Spraying/dusting is undertaken when there is no rainfall, soil moisture is 90% and wind speed is < 25 km/hr.

Crop monitoring: Meteorological tools such as crop growth models, water balance technique or remote sensing are used to evaluate the crop growth performance and recommend suitable remedial measures.

Crop modelling and yield-climate relationship: Suitable crop models can be developed by using the current and past weather data and these models can predict the growth and yield of a crop before its harvest.

Research on crop-climate relationship: Agrometeorology can help to understand crop-climate relationship so as to resolve complexities of a plant process in relation to its microclimate. Thus, the microclimatic conditions can be modified by appropriate agronomic practices for increasing agricultural productivity.

Climate extremities: Climatic hazards like frost, floods, droughts, hail storms, high wind velocity, *etc.* can be forecasted and the crops at various stages in the field may be saved by taking appropriate protective measures against these climatic extremities.

Climate as a tool to diagnose soil moisture stress: Soil moisture can be determined through water balance technique, which is used to diagnose the soil moisture stress and necessary protective measures such as irrigation, mulching, application of antitranspirant, defoliation, thinning, *etc.* can be undertaken to supply water to or reduce the moisture loss from the system through evapotranspiration.

Livestock production: Meteorological parameters can be related to the growth, development and production of livestock. Favourable and unfavourable weather conditions for livestock are studied in the agrometeorology and the climatic normal is worked out for optimisation of milk, meat and egg production. Animal growth simulation models may be developed for prediction of productivity in advance and also obtaining potential yield in different agroecological zones.

Fisheries: Fishermen need information on atmospheric and oceanic changes before they proceed on sea for fishing and this is possible from meteorological knowledge.

Weather-crop pest and disease relationships: The weather-crop pest and disease relationships are studied in agrometeorology. It helps in taking preventive and curative measures to avoid or check the outbreak and spread of biotic stress in crops and livestock.

Delineation of agroecological zones: Study of agrometeorology of a region helps to delineate agroecological zones for efficient use of resources and faster transfer of technology.

Crop weather calendars: Study of agrometeorology helps in preparing crop weather diagrams and crop weather calendars for timely agricultural operations by farmers.

Agroadvisory: Value added weather based agroadvisory can be developed for different agroclimatic zones and by its real time dissemination to farmers, the input cost can be reduced and productivity can be increased.

Soil formation: Soil formation process depends on climatic factors like temperature, precipitation, humidity, wind, *etc.*, thus climate is a major factor in soil formation and development.

Climate change: Study of agrometeorology helps to understand the impact of climatic variability or climate change on agricultural production systems. It also helps in devising suitable approaches to climate change adaptation and mitigation in agriculture.

Environmental pollutions: Modern intensive agriculture affects the environment through the generation of air pollutants, greenhouse gases (CO_2, methane and nitrous oxide), ammonia and tropospheric ozone. Other modes of production cause soil erosion by wind and water. Agrometeorology has a major role to play in the understanding of emissions and pollutions from various unsustainable production systems.

ATMOSPHERIC WEATHER VARIABLES

Weather of a place at a given instant of time is made up of multiple parameters known as weather variables or weather elements. Most weather elemements occur in the lowest layer of the atmosphere, i.e., the troposphere. The major weather variables are

- air temperature
- solar radiation
- air pressure
- humidity of air

- amount and kind of cloud cover
- amount and kind of precipitation
- speed and direction of the wind

Many factors affect the day-to-day weather and long-term climate of a given region. These include

- latitude (angle of sun's rays and effect on day length)
- altitude (height above sea-level)
- distance from the sea
- ocean currents (the movement of water at different temperatures through the oceans)
- mountains
- topography

Areas closer to the equator or sea level are generally warmer than areas far from the equator or at high altitudes. Local geography, including distance from the ocean and nearness to mountains, also has a significant influence. Even the type of vegetation in a region can exert a strong influence on local weather patterns.

APPLICATIONS OF AGROMETEOROLOGY

Despite impressive advances in agricultural technology, agricultural production is still dependent on weather and climate. The need for agrometeorological services have increased in recent years because of the increasing climate variability and associated extreme events. The fields of interest of agrometeorology include agricultural and horticultural production, forestry, animal husbandry, fisheries, agricultural planning, processing, transport and storage of food, poverty reduction and sustainable development aspects of the livelihood of farmers. Other important subjects are agroclimatic characterisation of pests and diseases and their safe control, quality of agricultural products, animal comfort aspects, plant production for other than food purposes, including biomass as a renewable energy resource and ecological considerations. Agrometeorological services are useful for the following activities.

- climate monitoring and analysis leading to planning and early warnings of climate anomalies
- real time monitoring leading to the provision of operational advices, for example, in determining irrigation efficiency

- microclimate manipulation and prediction, within crops, soils and managed environments (glass houses, stores, *etc.*)
- preparedness for agricultural hazards (including pests and diseases)

Rational use of climate information in agricultural production requires knowledge on climatic requirements of crops and climatic characteristics specific to a given farming area expressed in basic statistical terms. The crop-weather relationship could be better understood if the various aspects of environmental stress on crops are taken into account. Crop simulation modelling highlights the yield potential constraints analysis under different crop growing environments of a given crop. Rainfall, air and soil temperature, photosynthetically active radiation, soil moisture, wind, humidity and evapotranspiration are the atmospheric variables that play dominant roles in crop production. In developed countries mathematical models are increasingly used in operational agrometeorology, in conjunction with Geographic Information Systems to provide inputs to Decision Support Systems. These models have utilised meteorological observations. These forecasts may be exploited to increase the utility of models to decision makers. The possible utilities of agrometeorological information in agriculture are discussed below.

Monitoring and early warning

The availability of adequate and quantitative agrometeorological data is an absolute prerequisite for analysing and managing production processes in agriculture, including livestock and forestry. Decision making in agricultural operations for well grown crops or crops endangered by pests, diseases and/or other environmental disasters needs weather forecasting and climate prediction to the required accuracies. A number of crop monitoring systems and yield forecasts are now being implemented worldwide and upscaling to regional scales are being attempted. In these, commercial crops like soybean, maize, wheat, sorghum, *etc.* are continuously monitored and forecasted.

Crop, forest and livestock protection

Weather effects on plant, tree and animal discomfort and injury, as well as crop, forest and livestock losses are highly complex. Protection can take the preparative form of either planning crops, varieties and sites to avoid or mitigate the effects of the relevant meteorological extremes detrimental to plants or improving sites to reduce and/or avoid the impact of these extremes. Agrometeorology can play a significant role in reducing the negative impacts caused by pests and diseases. The use of climate in land-use planning, scheduling agricultural activities and preparing crop calendars as well as in crop, forest and livestock protection have benefited the agricultural communities of many regions. Information like return periods, frequencies of occurrences and intensities of extreme events and assured rainfall in different growing periods are valuable in choices for preparedness strategies. An appropriate, preferably integrated pest management system using meteorological and microclimatological information can reduce pre- and post-harvest losses appreciably. The tactical use of weather information in the prediction of pest and disease development allows for near optimum use and timing of pesticides and/or release of predators. Major applications of meteorology and climatology to forest operations include pest and disease control, frost protection and fire prevention. Monitoring fire-weather parameters such as moisture conditions of the soil and ground litter, wind speed and direction, temperature, humidity and lightning strikes are crucial for preparedness of forest fire and for early action before it gets out of control.

Water resources assessment

Because of the scarcity of water in many places of the world, prudent water management and increasing water use efficiency are essential, especially in arid and semiarid regions. Global and national policies and strategies are now being developed to increase awareness of water shortage, promote water conservation and water harvesting, redress mismanagement of ground water, increase water use efficiencies, promote the use of additional sources of water and promote recycling of water. In many countries, substantial resources have been used to monitor floods and droughts and to design appropriate irrigation and drainage systems for agriculture. Water budget/balance calculations, including elaborate evapotranspiration calculations based on physics and plant factors, and soil moisture determinations are widely applied tools in water waste determinations and water use planning in agriculture.

Soil and biomass resources assessment

Climate and weather affect the chemical, physical and mechanical properties of soil, the organisms it contains and its capacity for retaining and releasing heat and moisture. Rainfall adds chemical constituents to the soil but also washes out soil nutrients. Weathering is an important factor in determining the nature of soil. Topsoil composition, vegetation cover with surface contact and local weather factors largely determine the existence and extent of the problems of wind and water erosion. The state of the soil affecting cultivation, pest control and harvesting is also much influenced by weather conditions. Both chemical and physical components of soil degradation are of importance in soil resources assessments. Exposed agricultural land losing topsoil by wind and/or water actions needs to be stabilised and covered. It is important to keep the soil covered in economically useful ways. Demographic pressure on natural resources like forests has caused weather and climate to change locally, regionally and even globally. Forest fires are also a serious source of local and transboundary air pollution and atmospheric degradation. Correct application of meteorological and climatological information can be of considerable benefit to the protection, sustainable development and conservation of forest resources, greening of degraded areas through forest rehabilitation, afforestation and reforestation.

AGROMETEOROLOGY AND FUTURE NEEDS AND CHALLENGES

The welfare, safety and quality of human society are subject to profound effects of the vagaries of weather and climate. Meteorological hazards such as drought, flood, tropical storms, heat and cold waves may endanger lives and property and disrupt even the most sophisticated system. Climatic conditions have a complex bearing on soil, crops, plant disease and the factors, both biotic and abiotic that is related to agricultural and animal production systems. The greenhouse effect and global warming, destruction of ozone layer, protection of land and water resources and renewable energy are few areas where climate will be increasingly important.

Agroclimatological database and research

The availability of a proper meteorological and agrometeorological database is a major prerequisite for studying and managing the processes of agricultural production. A major and inevitable priority is to build a database of meteorological, phenological, soil related and agronomic information (Mavi and Tupper, 2004). The acquisition of pertinent climate and agrometeorological data, their processing, quality control, and archiving, and timely access and database management are important components that will make the information valuable to agrometeorological research and operational programmes (Stigter *et. al.*, 2000).

Crop-weather models for a large number of crops have been generated. But many of these models need to be further refined and tested before their practical application. Current research is being focused on detailed soil-water-crop relationships, determining the adjusted crop genetic coefficients, bridging simulation model outputs with user needs for applications, and developing practical decision support systems.

Early warning system and weather based agroadvisories

Agriculture plays a significant role in the overall socio-economic fabric of India. Weather becomes more significant in crop management. The farmers are adversely affected by climate risks in their farming, which include increase in temperature, decrease in rainy days, increase in precipitation intensity and amount, shorter winter periods, decreased ground water availability, increased occurrence of drought and floods, increased duration of water logging, *etc.* which in turn result in yield losses. Agrometeorology has advanced over decades from a descriptive to a quantitative science using physical and biological principles. Much of the developed world is less vulnerable to climate now-a-days than in the past. They have applied climate knowledge and early warning systems to allow adaption to mitigate the impacts of extreme events. Though India has acquired the relevant sophisticated technologies but these are yet to be fully translated to the benefits of the farming community.

Use of satellite remote sensing technology

Despite the technological advances, Indian farmers are mostly dependent on seasonal rains which are highly variable in time and space. Thus, weather forecasts are of great importance to agricultural activities. Much research has gone into characterisation of various crop environments, quantification of crop-weather interactions in relation to crop yield, crop weather modelling and crop-pest-weather dynamics. Now it is time to consolidate these findings in different locations, make use of this information and reach the farmers through agroadvisories on a larger scale in a more realistic way. These advisories are to be more farmer-friendly. A more accurate forecast along with actionable agroadvisories is required at block level. Remote sensing aided weather based agroadvisories may be a promising way which would enable the farmers to take the most appropriate actions on real time basis. Satellite remote sensing technology is increasingly gaining recognition as an important source of many agricultural applications as it is superior to the traditional methods in terms of accuracy and saving of time. Geographic Information Systems technology is becoming an essential tool for combining various maps and satellite information sources in models that simulate the interactions of complex natural systems.

Assessment of climate change

Agrometeorology has to play a leading role in the assessment of climate change, its impact on agriculture, and adaptation strategies to increasing climate variability. Future climate change and increased variability will complicate pest and disease management problems. This will require improved analyses of the weather to develop new pest management techniques and strategies.

Lot of research has been done in various aspects of agrometeorology and it is time to consolidate the results and translate them to actionable weather based agroadvisories. In addition to weather predictions, the agrometeorologists should be able to estimate the likelihood of unusual weather events and their potential impact on crops and animals through quantitative analysis, which will be a great challenge.

Chapter 2

Earth Atmosphere

The dynamic layer surrounding the earth above its surface containing various gases, moisture, aerosols, *etc.* is called atmosphere. Atmosphere can be defined as the gaseous envelope surrounding the earth. It is retained by earth's gravity.

The atmosphere protects life on earth by absorbing ultraviolet solar radiation, warming the surface through heat retention and reducing temperature extremes between day and night. The common name given to the atmospheric gases used in breathing and photosynthesis is air. The atmosphere has a mass of about 5.15×10^{18} kg, three quarters of which is within about 11 km of the surface. The atmosphere becomes thinner and thinner with increasing altitude, with no definite boundary between the atmosphere and outer space. The imaginary 'Karman line' lies at an altitude of 100 km above the earth's sea level, and commonly represents the boundary between the earth's atmosphere and outer space. The study of earth's atmosphere and its processes is called atmospheric science or 'aerology'.

The atmosphere serves the life-system on the earth by the following activities.

1. It fulfils the biological oxygen demand of the animal life.
2. It supplies the necessary precipitation or moisture.
3. It protects the biological life on the planet from harmful extraterrestrial radiations, by absorbing it though ozone.
4. It maintains the warmth of the earth through its green house effect, avoiding the temperature to fall to too extreme limits. The earth's temperature in the absence of atmosphere would have been +95° C (day), and –145° C (night).
5. It provides the necessary CO_2 which is basic input required for photosynthesis in plants to build biomass.
6. It provides the necessary medium for the transport of pollens, seeds, spores and insets.

7. Many physical, chemical and hydrological processes responsible for weather and climate occur in atmosphere only.
8. Atmosphere is a big reservoir of nitrogen. Some plants and microbes can fix this nitrogen for plant growth.
9. It conducts sound and diffuses heat and gases.
10. It supports combustion.

COMPOSITION OF THE ATMOSPHERE

Various gases, in varying quantities as well as the suspended solid and liquid particles called aerosols constitute the atmosphere. A gas consists of individual atoms or molecules that are separated, whereas a particle consists of aggregates of atoms or molecules bonded together. These gases and particles are bound to the earth by the gravitational attraction. Atmosphere can be thought of as a stage where weather processes take place.

The composition of atmosphere is not constant; it varies from time to time and from place to place. Most of the mass of the atmosphere is made up of gases that are thoroughly mixed and occur in constant proportions. They are called fixed gases. In addition, a range of gases called variable gases can have a variable concentration over time or space. Fixed gases make up more than 99 % of all gas molecules and occur in nearly constant proportion throughout the atmosphere's lowest 80 km. The most significant components of the atmospheric gases are nitrogen, oxygen and argon. By volume, dry air contains 78.084%, nitrogen, 20.946% oxygen, 0.934% argon, 0.038% carbon dioxide, and small amounts of other gases. Air also contains a variable amount of water vapour, on average around 1%.

Although both nitrogen and oxygen are essential to human life on the planet, they have little effect on weather and other atmospheric processes. The variable components, which make up far less than one per cent of the atmosphere, have a much greater influence on both short-term weather and long-term climate. Water vapour, CO_2, CH_4, N_2O, and SO_2 all have an important property. They absorb heat emitted by earth and thus warm the atmosphere, creating the 'greenhouse effect'. Though the greenhouse effect is sometimes portrayed as a bad thing, trace amounts of gases like CO_2 warm our planet's atmosphere enough to sustain life. Global warming, on the other hand, is a separate process that can be caused by increased amounts of greenhouse gases in the atmosphere. In addition to gases, the atmosphere also contains particulate matter such as dust, volcanic ash, rain, and snow. These are, of course, highly variable and are generally less persistent than gas concentrations, but they can sometimes remain in the atmosphere for relatively long periods of time.

Fixed gases

Nitrogen (N_2) gas is largely inert, meaning that it does not readily react with other substances to form new chemical compounds. It is an important constituent of many organic compounds. Its main function is to regulate combustion by diluting oxygen. It also indirectly helps in oxidation producing reactive nitrogen oxides. Oxygen (O_2) is required for the respiration of all life on earth, from humans to bacteria. It is extremely reactive and capable of combining with all other elements to form different compounds. Oxygen is also essential for most combustion and is consumed when any substance burns. It participates in oxidation. Argon (Ar) is a very inert noble gas, meaning that it does not take part in any chemical reaction under normal circumstances. Other fixed but inert gases present in trace concentration include neon, helium, krypton and xenon.

Table 2.1 Composition of the atmospheric gases (in dry atmosphere below 80 km)

Constituent	Symbol	% by volume	Total weight × 10^{17} kg
Nitrogen	N_2	78.08	38.65
Oxygen	O_2	20.95	11.84
Argon	Ar	0.93	0.66
Carbon dioxide*	CO_2	0.038	0.023
Neon	Ne	0.0018	0.00064
Helium	He	0.0005	0.00004
Methane*	CH_4	0.00017	0.00004
Krypton	Kr	0.00011	
Xenon	Xe	0.00009	
Ozone*	O_3	0.00006	
Hydrogen	H	0.00005	

* variable concentration

Variable gases

Gases whose volume mixing ratios change in time and space are variable gases. These gases are in very small amounts in the atmosphere. But they play a crucial role in the thermodynamics of the atmosphere. Carbon dioxide and water vapour are two vital atmospheric constituents.

Water vapour (H_2O) is contributed by evaporation from water bodies and soils and transpiration from plants. It is most abundant among the variable gases and accounts for about 0.25% of the total volume of the atmosphere. The mixing ratios of water vapour vary with location and time. While it is nearly absent in most of the atmosphere, its concentration can range up to 4%

in very warm, humid areas close to the surface. Despite its relative scarcity, atmospheric water probably has more impact on the earth than any of the major gases, aside from oxygen. Water vapour participates in the hydrological cycle, the process that moves water between the oceans, the land surface waters, the atmosphere, and the polar ice caps. When cooled sufficiently, water vapour forms clouds by condensing to liquid water droplets. Water vapour is also an important greenhouse gas. It is much more prevalent near the tropics than in the polar regions. The amount of water vapour constitutes the atmospheric moisture which regulates the weather process. Its concentration decreases rapidly with altitude and most atmospheric water vapour is found in the lowest 5 km of the atmosphere.

Carbon dioxide (CO_2) affects the earth's climate and plays a large support role in the biosphere. The carbon dioxide released by the animals during respiration is utilised by plants for photosynthesis which in turn release oxygen. The current atmospheric concentration of CO_2 is about 0.0386% or 386 ppm. It has increased steadily from 0.0280% in the mid 1800s, mainly due to anthropogenic combustions. Carbon dioxide is involved in a complex global cycle. It is released from the earth's interior and produced by respiration of biota, soil microbial, fuel combustion and oceanic evaporation. It is dissolved in the oceans and consumed by plant photosynthesis. The imbalance between emission and uptake leads to net increase in the atmosphere. It is projected that CO_2 concentration in the atmosphere by 2050 may be more than 450 ppm or 0.045%. This rate of increase in CO_2 concentration will lead to increase in the earth's temperature by 1.0 - 3.5° C by the turn of this century. Higher surface temperature will lead to more melting of snow and ice, thereby causing sea level rise, a great threat to the coastal areas.

Ozone (O_3) is another variable gas, which is produced through the breakup of oxygen molecules in the upper atmosphere by solar ultraviolet (UV) radiation and is destroyed by reactions involving nitrogen oxides and chlorine. The ozone molecule is made up of 3 atoms of oxygen. Ozone gas is irritating to peoples' eyes and skin, and chemically attacks rubber and plant tissue. Fortunately ozone occurs only in minute amounts *i.e.*, 0.15 ppm in highly polluted urban air. Almost all ozone is found in a layer about 15 - 60 km in altitude, where their concentration can be up to 15 ppm. They are called stratospheric ozone and are vital to life on earth. Stratospheric ozone absorbs most of the high energy UV radiation from the sun that is harmful to plants and animals.

Methane (CH_4) is another variable gas whose present concentration in the atmosphere is about 1.775 ppm. Its concentration has been doubled in the last 200 years and most of the increase is attributed to low land rice cultivation and fossil fuel extraction. Despite its low concentration, methane is extremely effective absorber of thermal radiation emitted by earth's surface and has a crucial role in warming of the atmosphere.

Other variable gases include carbon monoxide, sulphur dioxide, nitric oxide, nitrogen dioxide and chloroflurocarbons (CFCs). They are contributed to the atmosphere from both natural and anthropogenic sources.

Aerosols

In addition to gases, the atmosphere has a wide variety of tiny particles suspended in the air, known collectively as aerosols. These particles may be liquid or solid and are so small that they may require very long times to settle out of the atmosphere by gravity. Aerosols include bits of suspended soil or desert sand, tiny smoke particles from a forest fire, salt particles left over after a droplet of ocean water has evaporated, plant pollen, very fine seeds of plants, spores, bacteria, volcanic ashes, and particles formed from the pollution created by a coal burning power plant. Aerosols normally occur at concentrations of about 10,000 particles per cm^3.

Aerosols are of various sizes. Raindrops suspended in a cloud are about 1–6 mm in diameter. Fine desert sand and cloud droplets are about 0.01 mm. Sea salt particles and smoke particles are about 0.0001 mm in diameter. Aerosols significantly affect the atmospheric heat balance, cloud growth, and optical properties. Some aerosols are just the right size to efficiently scatter sunlight, making the atmosphere look hazy. Under the right conditions, aerosols act as collecting points for water vapour molecules, encouraging the growth of cloud droplets and speeding the formation of clouds. They may also play a role in earth's climate. The aerosols are known to reflect a portion of incoming solar radiation back to space, which lowers the temperature of the earth's surface. Urban smog and dust storms which include aerosols can reduce visibility significantly.

VERTICAL THERMAL STRUCTURE OF ATMOSPHERE

There is no sharp boundary between the atmosphere and outer space. The earth's atmosphere can be divided into zones or layers arranged like spherical shells according to altitude above the earth's surface. The structure is defined in terms of the variation in pressure, density and temperature with height. In general, air pressure and density decrease exponentially with increasing altitude in the atmosphere. However, temperature has a more complicated profile with altitude, and is controlled by the distribution of gases that absorb ultraviolet and thermal infrared radiation.

In 1962 World Meteorological Organisation (WMO) divided the atmosphere into four regions based on temperature changes with height. From bottom to top, they are troposphere, stratosphere, mesosphere and thermosphere. This is called as the thermal structure of atmosphere. The temperature profile of the atmosphere is given in Fig 2.1.

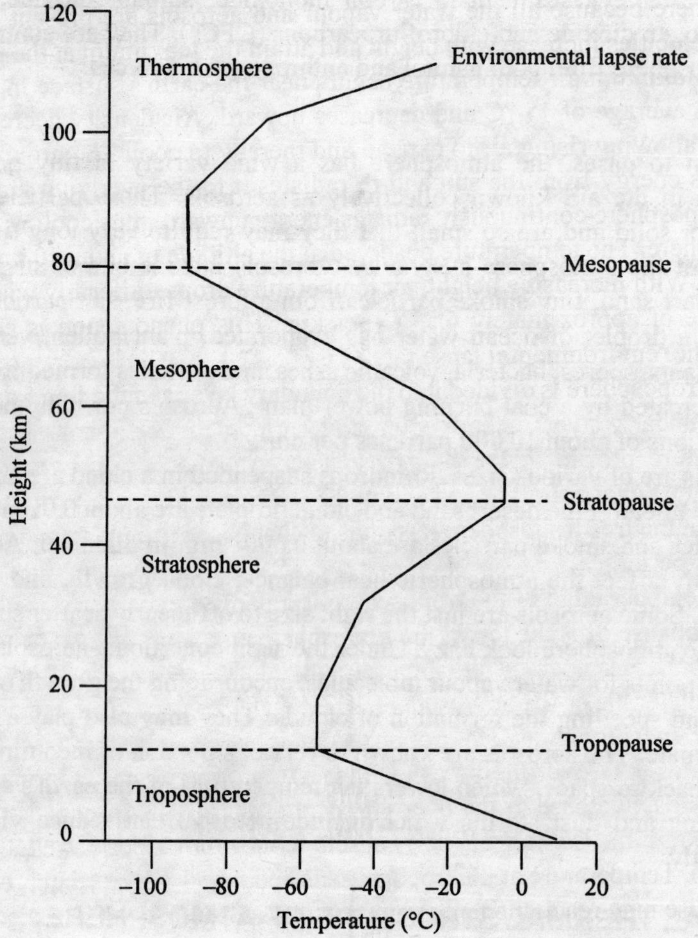

Fig 2.1 The temperature profile of the atmosphere

Troposphere

The troposphere is the lowest layer of earth's atmosphere. The word 'tropo' means mixing or turbulence and 'sphere' means region. Thus, 'troposphere' means the realm of mixing. In the troposphere, the air is vigorously mixed and stirred by storms, convection, and wind systems. The thickness of this layer varies from about 8 to 18 kilometres and the average thickness is approximately 11 kilometres. Greatest depths occur at the tropics where warm temperatures cause vertical expansion of the lower atmosphere. From the tropics to the earth's polar regions the troposphere becomes gradually thinner.

The troposphere contains roughly 80% of the total mass of the atmosphere. Various types of clouds, thunderstorms, cyclones and anticyclones occur in this sphere because all the water vapour and aerosols are present in it. The wind velocities increase with height and attain the maximum at the top of this layer. Maximum air temperature occurs near the earth's surface in this layer with an average of 15 °C and decreases upward. Air density decreases with height, allowing rising air to expand and thereby to cool. The ground surface receives energy from the sun and re-radiates it to the overlying atmosphere. The troposphere continuously radiates energy upward, thus cooling the upper troposphere. The temperature observed at the altitude of about 10 km is about –50° C. With increasing height, air temperature drops uniformly with altitude at a rate of approximately 6.5 °C per km. This phenomenon is commonly called the 'environmental lapse rate'.

The troposphere is divided into the boundary layer and the free troposphere. The boundary layer extends from the surface to 0.5–3 km altitude. The temperature profile in the boundary layer responds to changes in ground temperature over a short period of time (less than an hour), whereas temperature profile in the free troposphere responds to changes in ground temperature over a longer period.

At the upper edge of the troposphere is a narrow transition zone known as the tropopause. It is the top of the troposphere and the troposphere stops here. The tropopause separates the intensely mixed air of the troposphere from the much quieter air of the stratosphere. The height of tropopause is not constant in either space or time. The tropopause heights are higher (15–18 km) over the equator than over the poles (8–10 km). Strong vertical motions over the equator raise the base of the overlying stratosphere. The tropopause could be described as an 'isothermal' belt *i.e.*, the temperature does not change as we move up. Temperature at the tropopause is about –50 °C. In winter, equatorial tropopause may reach the temperature as low as –80 °C.

Stratosphere

This layer exists above the tropopause and extends to altitudes of about 50–55 km. The stratosphere contains about 19.9% of the total mass of the atmosphere. In the first 9 km of the stratosphere, temperature remains constant at about –57 °C. From an altitude of 20 to about 32 km, the temperature rises gradually with increase in altitude and then rises rapidly till the top of the stratosphere. The higher temperatures found in this region of the stratosphere is due to a localised concentration of ozone gas molecules. The stratosphere is home to the ozone layer. In this layer, the chemical bonds between oxygen atoms in molecular oxygen break and when the molecules absorb ultraviolet radiation the bond in ozone (O_3) breaks. Re-forming those bonds releases heat energy so the temperature increases with altitude in this layer. This layer of ozone is also called the ozone layer. The ozone layer is important to organisms at the earth's surface as it protects them from the harmful effects of the sun's ultraviolet radiation. Without the ozone layer life could not exist on the earth's surface. The ozone layer is not composed primarily of ozone. The peak concentration of ozone detected at 25 km above the sea level is only 10 ppm. The top of the stratosphere is called the stratopause, which is an isothermal belt like the tropopause. The maximum temperature in the stratosphere occurs at the stratopause and may exceed 0 °C. There is not much convection in the stratosphere because it has low temperature at the bottom than at the top. A very little weather phenomena may occur in the stratosphere.

Mesosphere

This layer extends from the stratopause at an altitude of about 50 km to the mesopause at 80–85 km above sea level. Temperatures drop with increasing altitude to the mesopause that marks the top of this middle layer of the atmosphere. It is believed that absorption of solar radiation near the base of the mesosphere provides most of the heat for the layer, which is dispersed upward by vertical air motions. The temperature gets down to nearly –95 °C at the mesopause, which tops the mesosphere. This is the coldest layer in the atmosphere. The mesosphere can be thought of as a resumption of the troposphere after the rude interruption of the ozone layer in the stratosphere.

Thermosphere

This layer lies beyond mesosphere at a height of about 80 km above the earth's surface and extends up to 500–1000 km. Thermosphere is the last thermally defined atmospheric layer and has no well defined upper limit. This atmospheric layer undergoes a gradual increase in temperature with height. Unlike the stratosphere, wherein a temperature inversion is due to the absorption of radiation by ozone, the inversion in the thermosphere occurs due to the extremely low density of its molecules. The temperature of this layer can rise as high as 1500 °C, though the gas molecules are so far apart that its temperature in the usual sense is not very meaningful. Even though the thermosphere has a very high proportion of molecules with immense amounts of energy, the thermosphere would still feel extremely cold to a human in direct contact because the total energy of its relatively few number of molecules is incapable of transferring an adequate amount of energy to the skin of a human. In other words, a person would not feel warm because of the thermosphere's extremely low pressure. This layer is completely cloudless and free of water vapour.

VERTICAL COMPOSITION OF ATMOSPHERE

The atmosphere can be divided into two spheres on the basis of its chemical composition occurring with height, viz., homosphere and heterosphere.

Homosphere

In the lower region up to the height of 80 km the various gases are thoroughly mixed and are homogenous by the process of turbulent mixing and diffusion. The bottom three thermal layers of the atmosphere (troposphere, stratosphere and mesosphere) together constitute the homosphere. The composition of the atmosphere is almost constant, except for the water vapour, carbon dioxide and ozone in homosphere. By the mixing mechanisms, component gases move from one part of the atmosphere to another and results in a fairly uniform chemical composition at different levels. Ionisation of the gases are almost absent in the homosphere. A greater concentration of ozone is found in the stratosphere, but ozone is a non-ionised gas. Throughout the homosphere, nitrogen and oxygen gases dominate.

Heterosphere

The composition of the atmosphere is not homogenous above 80 km. Because of the heterogenous composition, this upper atmosphere is called heterosphere. Varying mixes of gas molecules and ions are found in heterosphere. In heterosphere the main gases stratify by their molecular weights and are arranged into four roughly spherical shells, each of which has a definite chemical composition: nitrogen layer, oxygen layer, helium layer and hydrogen layer.

OTHER VERTICAL LAYERS OF ATMOSPHERE

Within the principal layers which are largely determined by temperature, several secondary layers may be distinguished by other properties.

Exosphere

The outermost layer of the atmosphere is sometimes called as exosphere. It lies beyond about 500 km and is characterised by increasing hydrogen and helium content. The oxygen and nitrogen that dominate the lower atmosphere dissociate into ions in the ionosphere. Hydrogen and helium molecules can and do easily leave earth orbit for outer space from the top of the exosphere. At this altitude, when their trajectories are changed by collision with another molecule to a generally upward direction, they have less and less chance of colliding with another one and being bounced back earthward. So, there is a constant loss of H and He to outer space from the upper exosphere, and earth's gravity is not strong enough to hold onto these light gases.

Ozone layer

The ozone layer is present within the stratosphere. In this layer, ozone concentrations are about 2 to 8 parts per million, which is much higher than in the lower atmosphere but still very small compared to the main components of the atmosphere. It is mainly located in the lower portion of the stratosphere from about 15 to 35 km, though the thickness varies seasonally and geographically. About 90% of the ozone in our atmosphere is contained in the stratosphere.

Ionosphere

Ionosphere is an atmospheric layer defined based on its electrical properties. This layer extends from the upper mesosphere into thermosphere and contains large number of electrically charged particles called ions. Although ionisation occurs at heights as high as 1000 km and extends as low as 50 km, the positive and negative charged ions are most dense in the range of 80 to 400 km. It is the first line of defence for earth against extremely shortwave radiation and high energy particles from the sun and cosmic rays. These particles are ionised atoms that include isolated protons and alpha particles. Cosmic rays are hazardous to life on earth. These rays and high energy fast moving particles smash into the few molecules of the ionosphere with such force that they strip them of electrons, turning them into ions, or electrically imbalanced atoms. The ions, with their electrical imbalances, are drawn by the earth's magnetic field and align themselves with that field's lines of force. This means that they are concentrated in great abundance where those magnetic field lines converge, at the north and south magnetic poles. The ionosphere forms the inner edge of the magnetosphere. The ionosphere is important to radio communications since radio waves can bounce off the layer and travel great distances. It has practical importance for radio propagation on earth. The ionosphere consists of three layers of varying ion density. From bottom to top these layers are called the D, E and F layers.

Planetary boundary layer

The planetary boundary layer is the part of the troposphere that is closest to earth's surface and is directly affected by it, mainly through turbulent diffusion. During the day the planetary boundary layer usually is well-mixed, whereas at night it becomes stably stratified with weak or intermittent mixing. The depth of the planetary boundary layer ranges from as little as about 100 metres on clear, calm nights to 3000 m or more during the afternoon in dry regions.

Chapter 3

Atmospheric Pressure

Atmospheric pressure is the force per unit area exerted on a surface by the weight of air above that surface in the atmosphere of earth. The gas molecules of the atmosphere are pulled towards the earth by the force of gravity and hence exert a force on earth surface. In most circumstances atmospheric pressure is closely approximated by the hydrostatic pressure caused by the weight of air above the measurement point. Atmospheric pressure influences other weather elements in a significant way. Small changes in air pressure may not be felt, but they are important for bringing a change in weather. Atmospheric pressure can be calculated as the summed weight of all gas molecules between a horizontal plane and top of the atmosphere and divided by the area of the plane. Low pressure areas have less atmospheric mass above their location, whereas high pressure areas have more atmospheric mass above their location.

FACTORS AFFECTING ATMOSPHERIC PRESSURE

Air pressure never remains constant. Air pressure is greatly influenced by air temperature, altitude, water vapour presence in the air, rotation of earth, *etc*. However, these factors are not independent of each other in affecting the atmospheric pressure.

Temperature: Hot air expands and exerts low pressure. Cold air contracts and exerts high pressure. Thus, increase in temperature decreases the air pressure. Places nearer to equator experience low pressure due to prevalence of high temperature. The air being cold and heavy throughout the year a high pressure belt is created in both polar regions.

Altitude: Pressure varies smoothly from the earth's surface to the top of the mesosphere. The pressure depends on the density or mass of the air. The density of air depends on its temperature. The density of air decreases with increase in height so the pressure also decreases with increase in height. At low altitudes above the sea level, the pressure decreases by about 1.2 kPa for every 100 metres.

Water vapour: Water vapour content of air is low in cold areas than that in dry areas. Hence, moist air of high temperature in dry areas exerts less pressure compared to cold air.

Time: Air pressure also varies over time, and these temporal differences are usually caused by the temperature of the air. Cool air is denser (heavier) than warm air. Warm air is less dense (lighter) than cool air and will therefore rise above it. Areas of high pressure can be caused when cool air is sinking and pressing on the ground. At this time, the weather is usually dry and clear. In contrast, when warm air rises, it causes a region of low pressure. With low pressure, the weather is often wet and cloudy.

Rotation and gravity of the earth: Due to rotation of the earth, pressure at 60 to 70 degree latitude in both the hemispheres becomes low. Near sub-polar belts, the earth's rotation makes the air to escape from these belts, which move towards the horse latitude (30–35° N and S). These latitudes absorb the air from sub-polar belts making the pressure high. Gravity shapes and influences all atmospheric processes. It causes the density and pressure of air to decrease exponentially as one moves away from the surface of the earth.

VERTICAL AIR PRESSURE VARIATION

As elevation increases there is less overlying atmospheric mass, so that pressure decreases with increasing elevation. Pressure varies smoothly from the earth's surface to the top of the mesosphere. The elevation is often expressed as geopotential height rather than geometric height in weather studies. Geopotential height is a vertical coordinate referenced to earth's mean sea level and an adjustment to geometric height (elevation above mean sea level) using the variation of gravity with latitude and elevation. Thus, it can be considered as a 'gravity-adjusted height'.

The International Standard Atmosphere (ISA) is an atmospheric model of how the pressure, temperature, density, and viscosity of the earth's atmosphere change over a wide range of altitudes. It consists of tables of values at various altitudes, plus some formulae by which those values were derived. The ISA model divides the atmosphere into layers with linear temperature distributions. The other values are computed from basic physical constants and relationships. The model is based on average conditions at mid-latitudes. The temperature and pressure values in the standard atmospheric layers are given in table 3.1. In the table geopotential height is calculated from a mathematical model in which the acceleration due to gravity is assumed constant.

Table 3.1 Temperature and pressure values in standard atmospheric layers

Layers	Level name	Geopotential height (km)	Geometric height (km)	Lapse rate (°C/km)	Temperature (°C)	Pressure (mb)
0	Tropo-sphere	0.0	0.0	−6.5	+15.0	1013.25
1	Tropopause	11.000	11.019	+0.0	−56.5	226.32
2	Strato-sphere	20.000	20.063	+1.0	−56.5	54.75
3	Strato-sphere	32.000	32.162	+2.8	−44.5	8.68
4	Stratopause	47.000	47.350	+0.0	−2.5	1.11
5	Mesosphere	51.000	51.413	−2.8	−2.5	0.67
6	Mesosphere	71.000	71.802	−2.0	−58.5	0.039
7	Mesopause	84.852	86.000		−86.2	0.004

About 50% of atmospheric pressure from the mean sea level decreases within 5 kilometres of height from the earth's surface. However, pressure decrease with altitude is not constant. It decreases at a much faster rate near the earth surface where the pressure is relatively more than that at top where air is less dense and pressure is relatively less. Table 3.2 describes the pressure variation with altitude in the lower atmosphere.

Table 3.2 Atmospheric pressure change with altitude

Height (km)	Atmospheric pressure (mb)	Change in atmospheric pressure compared to MSL (%)
MSL	1013	–
5	540	53.3
10	265	26.1
15	121	11.9
20	55	5.3
25	25	2.3
30	12	1.1

Atmospheric pressure in the upper atmosphere is depicted through height contours instead of isobars. The upper air charts typically show the plot of geopotential height for individual pressure level such as 850, 700, 500 and 300 mb corresponding to conditions at about 1500, 3000, 5500 and 10000 m above sea level, respectively. These are surfaces having equal atmospheric pressure and may be seen as troughs (lows) or ridges (highs). An atmosphere in which isobaric surfaces are also surfaces of constant density are known as 'barotrophic atmosphere'. It is a state of atmosphere where surfaces of constant pressure and constant density coincide, so that the geostrophic wind is independent of height. The atmosphere which is not barotrophic is called 'baroclinic atmosphere'. In such atmospheric state, density depends upon both temperature and pressure and the geostrophic wind varies with height and is related to the horizontal temperature gradient.

HORIZONTAL AIR PRESSURE VARIATION

Although vertical air pressure variation and distribution are important, meteorologists are more interested in pressure differences that occur horizontally. Isobars are used to show pressure distribution. An isobar is a line connecting points with equal values of pressure. The rate of change in atmospheric pressure between two points at the same elevation is called pressure gradient or isobaric slope. It is proportional to the difference in pressure and is the immediate cause of horizontal air movement. The direction of airflow is from high to low pressure and the speed of flow is directly related to pressure gradient. The pressure gradient is steep when the rate of change is great, and the steeper the gradient, the more rapid will be the flow of air.

Factors which govern the distribution of temperature on the earth's surface equally govern the distribution of pressure as well. The equatorial region which receives maximum amount of insolation has high temperature and low pressure. Contrary to this, the polar regions with their low temperature develop areas of high pressure. Thus, a low pressure belt in the vicinity of equator and a high pressure belt at and near the poles are developed and there is gradual increase in surface air pressure from the equator towards the poles. However, some of the air rising near the equator and moving towards the poles in the upper atmosphere is turned aside by earth rotation or by other dynamic processes. They give rise to subtropical high pressure belt at about 30 to 40 degree latitude and sub-polar low pressure belt at about 50 to 65 degree latitude in both the hemisphere. Thus, there are in all seven alternating low and high pressure belts on the earth's surface; equatorial trough of low pressure, subtropical high pressure belt (northern hemisphere), subtropical high pressure belt (southern hemisphere), sub-polar low pressure belt (northern hemisphere), sub-polar low pressure belt (southern hemisphere), polar high (northern hemisphere) and polar high (southern hemisphere).

Occurrences of high and low pressures do not always depend on geographical location. This makes isobars on weather maps meaningful and useful tools. In a daily weather chart, isobars display characteristic shapes, which indicate different types of pressure systems. A pressure system is a region of the earth's atmosphere where air pressure is unusually high or low. High and low pressures develop constantly due to temperature differentials in the atmosphere and water of oceans and lakes. When the isobars are circular or elliptical in shape, and the pressure is the lowest at the centre, such a pressure system is called 'low' or 'depression' or 'cyclone'. A line of low pressure is called a 'trough'. The word 'cyclone' is derived from a Greek word '*cyclones*' meaning the coils of a snake. The diameter of a cyclone ranges from a few hundreds to 2000 kilometres. Low pressure systems are generally formed by air becoming less dense as a result of heating. This is generally the reason for storms and tropical cyclones.

When the pressure is the highest at the centre, the pressure system is called a 'high' or 'anticyclone'. Anticyclones are never as intense as the cyclones, but are generally larger in size than the cyclones. The isobars are usually more symmetrical and circular. Around high-pressure systems, winds flow clockwise in the northern hemisphere, anticlockwise in the southern hemisphere. When the isobars are elliptical rather than circular, the system is called a 'ridge' or 'wedge'. The pressure gradient is directed from the centre towards the outer rim. They occur often, but not always, when there is an uneven heating of the ground, causing the hot air to rise and spread out. This rising air creates an air 'void', or a low-pressure system in the area above the ground. However, when the air cools and sinks back down to the ground, a high-pressure system is created.

DIURNAL AND SEASONAL PRESSURE VARIATION

Insolational warming and radiational cooling are the principal reasons for diurnal variation in air pressure. Atmospheric pressure shows a diurnal (twice-daily) cycle caused by global atmospheric tide. These variations have two superimposed cycles, a circadian (24 hour) cycle and semi-circadian (12 hour) cycle. At the time of equinoxes, the two maxima of pressure are recorded at 10 a.m. and 10 p.m., while the two minima are observed at 4 a.m. and 4 p.m. This effect is the strongest in tropical zones and almost zero in polar areas. In the same latitude, the places situated nearly at sea-level record relatively larger amount of variation in daily pressure.

Distinct seasonal variations in atmospheric pressure are also observed at a given locality. This is clearly the effect of annual variation of insolation received. Over the continents, the high pressures are generally recorded during cold season, while over the oceans they are observed during the warm season. Equatorial regions record the smallest amount of variation in their seasonal pressure as there is little variation in insolation throughout the year. In high latitudes, seasonal variation in pressure tends to be much large because of an increasing contrast in angle of incidence and the length of daylight.

Chapter 4

Wind and its Types

A big layer of air called the atmosphere surrounds the earth. The air within this layer moves from place to place when it warms up or cools down. This moving air is known as wind. In other words, air in motion is called wind. Winds move moisture and heat around the world and also produce much of our weather.

The strength of the wind can vary enormously. Sometimes air moves slowly and the wind is barely noticeable. When the weather is clear there may be a gentle breeze. When the wind is still very light, but it can be felt and there may be rustlings by leaves. At other times, the air can move very quickly and become a gale or hurricane, blowing down trees and damaging electric poles and buildings.

Winds are part of a global air circulation system that acts to balance temperature and pressure around the world. Different parts of the world receive different amounts of heat from the sun. This differential heating in turn results in differences in temperature and air pressure around the world, which drives the world's winds.

As equatorial areas are heated most, the air above them warms and rises as it becomes lighter than the surrounding air, causing an area of low pressure. In cooler areas, the air sinks because it is heavier and results in an area of high pressure. Winds will blow as air is squashed out by the sinking cold air and drawn in under the rising warm air. Any difference in temperature like this will always cause a difference in air pressure, and therefore winds will blow. These movements result in a global wind pattern, with air moving between different areas around the world and also at different heights in the atmosphere. Colder air from the poles tends to sink and move towards the equator closer to the surface of the earth. In contrast, warm air from the equator rises and moves towards the poles high in the atmosphere because it is lighter.

WIND DIRECTION AND SPEED

The atmospheric motion is resulted from the difference in air pressure. On both local and global scale, air has a tendency to move from higher to lower air pressure zone. Air pressure exerts a force on air molecules and this can control air motion. There are two types of movement in the atmosphere, horizontal movement and vertical movement. The term wind generally refers to the horizontal movement of air relative to the earth's surface. Vertically moving air columns are called as currents. The air moves vertically when it is warmer and consequently more buoyant than the surrounding air. Upward and downward air currents are referred to as updrafts and downdrafts, respectively. Vertical motion in the atmosphere is of significance for the formation of clouds, precipitation and various types of storms. However, more air is involved in the horizontal movement than the vertical movement. The horizontal differences in air pressure are far more variable than vertical changes and give rise to a wide range of air motions. Winds are the means by which the uneven distribution of pressure over the globe is balanced out. Winds also act as a means of transporting heat, moisture and other properties from one part of the earth to another.

The wind is characterised by its direction and speed. Wind velocity is a vector quantity which has both magnitude and direction. The magnitude of wind is called wind speed and is a scalar quantity. Wind direction is reported by the direction from which it is blowing. For example, a northerly wind blows from the north to the south. Wind speed is the rate of movement of air in its instantaneous direction. Wind direction is determined with reference to true north and is expressed in degrees measured clockwise from the north. For example, a wind blowing from north is represented as 0° or 360°. The east, south, west winds are denoted as 90°, 180° and 270°, respectively. When the wind consistently blows more often from one direction than from any other, it is called a prevailing wind.

FACTORS AFFECTING SPEED AND DIRECTION OF WIND

There are four controls on the horizontal movement of air near the earth's surface; the pressure gradient force, the Coriolis force, centrifugal force and frictional force.

Pressure gradient force

The wind is caused by differences in pressure. When high pressure exists in one location and low pressure exists nearby, air moves from high to low pressure. The force causing this motion is the pressure gradient force. The pressure gradient force has both vertical and horizontal components. The vertical pressure gradient force act equally in the opposite direction of the gravitational pull of the atmosphere and provide the hydrostatic balance. The vertical air motion is limited by this hydrostatic equilibrium and vertical air motion becomes possible when the equilibrium is disturbed.

The horizontal pressure gradient is developed through spatial differences in surface heating and consequent changes in air density and pressure. The speed of flow is directly related to the pressure gradient. The pressure gradient force per unit mass is expressed as, $f = -1/\rho * (\partial p/\partial n)$ where, f is the pressure gradient force, ρ is the air density and ∂p is the pressure change over distance ∂n. Thus, $\partial p/\partial n$ is the pressure gradient. The negative sign indicates that the force operates from high to low pressure; p decreases as n increases. The closer the isobar spacing, the more intense is the pressure gradient and greater the wind speed. The pressure gradient force is also inversely proportional to the air density.

Coriolis force

In the rotating earth, there is apparent deflection of all free moving objects including wind to the right of their line o. motion in the northern hemisphere and to the left in the southern hemisphere as viewed by observers on the earth. A French mathematician G. D. Coriolis postulated this effect in 1844 and it is known as Coriolis force. This force is expressed as, $f = 2\omega V \sin\theta$; where, f is the Coriolis force, ω is angular velocity of spin (15°/hour for the earth), V is the velocity of the mass and θ is the latitude. $2\omega V\sin\theta$ is also referred as the Coriolis parameter. The magnitude of the deflection is directly proportional to the horizontal velocity of the air and the sine of the latitude ($\sin 0 = 0$ and $\sin 90 = 1$). The effect is thus a maximum at the poles and becomes zero at the equator. It affects the direction but not the speed of the air motion. The Coriolis force always acts at right angles to the direction of the air motion, to the right in the northern hemisphere and to the left in the southern hemisphere. This effect is also called as the earth's rotational deflective force.

Centrifugal force

When a body follows a curved path, it experiences an inward acceleration towards the centre of rotation. In the rotation of the air about a local axis of high or low pressure, the curved path of the air (parallel to the isobars) is maintained by an inward-acting or centripetal acceleration. The centripetal force, however, is opposed by an equal and opposite centrifugal force acting radially outward from the centre of curvature. The magnitude of the centripetal acceleration is generally small, but it becomes important where high velocity winds are moving in much curved paths.

A closed isobaric system with the lowest isobar at the centre gives rise to cyclones. If the highest pressure is at the centre and the lowest at periphery, the system represents anticyclone. In a cyclone, wind blows from margins to the centre, anticlockwise in northern hemisphere and clockwise in the southern hemisphere. In anticyclones, wind blows from centre towards margins, clockwise in northern and anticlockwise in the southern hemisphere.

Frictional forces

The roughness of the surface of the earth provides frictional resistance to the air motion. It is the retarding effect of trees, buildings and other irregularities in the topography. Thus, near the earth's surface, friction causes the wind to be slower than it would be otherwise. The layer of frictional influence is known as the 'planetary boundary layer'. Surface friction also causes winds to blow more inward into low pressure areas. It acts exactly in the opposite direction from the wind. The influence of frictional force diminishes with increasing elevation. The frictional force over oceans and deserts is small, whereas it is large over forests and buildings. On an average, surface friction turns winds 15° to 45° towards low pressure, with lower values corresponding to smooth surfaces and higher values corresponding to rough surfaces. The slowing of the wind towards the surface modifies the Coriolis force, which is dependent on velocity, causing it to decrease. The height above ground where surface friction has a negligible effect on wind speed is called the 'gradient height' and the wind speed above this height is assumed to be a constant called the 'gradient wind speed'.

TYPES OF WINDS ON SURFACE AND ALOFT

Geostrophic wind

In the upper atmosphere, the effects of friction on airflow are very small and can be disregarded. Here, the wind involves only the pressure gradient force and the Coriolis force. Under the idealised conditions, when the pressure gradient force is exactly equal and opposite of the Coriolis force, the airflow is said to be in geostrophic balance. The wind generated by this balance is called geostrophic wind. Geostrophic wind arises above the boundary layer, where surface friction is negligible and along straight isobars. If there is a horizontal pressure gradient, represented by two parallel isobars, the pressure gradient force will cause still air to move from high to low pressure. As the air moves, the Coriolis force deflects the air to the right. The Coriolis force continues deflecting the air until the force exactly balances the magnitude and direction of the pressure gradient force. The resulting wind is the geostrophic wind and it flows parallel to isobars

Gradient wind

Unlike geostrophic winds that flow in a straight path, winds around cells of high or low pressure follow curved paths. Winds of this nature which blow at constant speed parallel to curved isobars are called gradient winds. Here, in addition to pressure gradient force and apparent Coriolis force, centrifugal force (or centripetal acceleration) also comes into picture. When centre of low and high pressure exist aloft, the wind is controlled by pressure gradient, apparent Coriolis and centrifugal forces. For a low pressure centre, the pressure gradient force acts towards the centre of low whereas centrifugal force and Coriolis force act away from the centre of low. Geostrophic wind is a special case of gradient wind in which the centrifugal force is negligible as compared to the pressure gradient force and Coriolis force.

Thermal wind

Wind aloft at any level is considered to have two components (i) Geostrophic wind above the friction layer and (ii) a wind blowing along the mean isotherms with low temperature to its left in the northern hemisphere and to its right in the southern hemisphere. This second component is called thermal wind. It is the difference in the geostrophic wind between two levels in the atmosphere. It exists only in an atmosphere with horizontal temperature gradiants. Thermal wind is a fictitious wind and it has no actual existence. However, the concept is of great importance in practical weather prognosis. The thermal wind concept explains how differences in wind speed with height are dependent on horizontal temperature differences and explains the existence of the jet stream.

Vertical motion and vorticity

Mass uplift or descent of air occurs primarily in response to dynamic factors related to horizontal airflow. Air may accelerate or decelerate leading to velocity divergence or convergence, respectively. Convergence or divergence may also occur as a result of frictional effects. Convergence or divergence gives rise to vertical stretching or sinking. Air rises above a low pressure cell and subsides over a high pressure one, with compensating divergence and convergence, respectively in the upper troposphere. The rotation or angular velocity of small parcels of air is its 'vorticity'. The air within a low pressure system may be regarded as comprising an infinite number of small air parcels, each rotating cyclonically about an axis vertical to the earth's surface. Vorticity has three elements; magnitude, direction and the sense of rotation. If all winds are geostrophic, there could be no convergence or divergence and hence no weather.

SURFACE WIND SYSTEMS

Unequal heating of the earth's surface generates pressure gradient which gives rise to wind. There are three latitudinal circulations and there are also important longitudinal variations around each hemisphere.

Trade winds

The condition of the greatest heating and expansion at the equator causes rising of air and creating low pressure belt at 5° N to 5° S latitude known as doldrums or equatorial low or calm. The rising of air from equator causes increase in air pressure at 35° N and 35° S latitude which is known as sub-tropical high or horse latitude belt. The winds therefore flow from horse latitude to the equatorial region and called as 'trade winds'. While moving, these winds are deflected by Coriolis force to the right in the northern hemisphere and to the left in southern hemisphere and become North-East trades and South-East trades in northern and southern hemispheres, respectively. The blow of air from equator and accumulation of air over 25–35° latitudes giving rise to high pressure belt region of descending air is known as Hadley cell.

Westerlies

The low pressure area at about 60–65° latitudes in both the hemisphere is known as sub-polar low or polar front. The winds that flow from sub-tropical high pressure area at 25–35° latitude in both the hemisphere to the low pressure area, situated at about 60–65° latitude in both the hemisphere, are known as anti-trade winds. In the upper atmosphere the reverse air movement takes place. This circulation is known as Ferrel cell. These winds instead of flowing in straight line are deflected due to Coriolis force. In northern hemisphere their direction is north-west and in southern hemisphere it is south-west, thus named as westerlies. A very strong westerly wind which blows almost continuously in the southern hemisphere at a latitude of 40° is called 'roaring forties'

Polar easterlies

A permanent high pressure area exists near the poles due to shrinkage and cooling of air. Therefore winds flow from the polar high to sub-polar low pressure area at about 60–65° latitude. Due to Coriolis force, winds flow in north-east direction in northern hemisphere and in south-east direction in southern hemisphere, thus, named as polar easterlies. These winds consist of cold air. The air circulation is known as polar cell.

CYCLONES AND ANTICYCLONES

A cyclone and its counterpart, an anticyclone is a large-scale system of air circulation in the atmosphere in the zones between the equator and either of the poles.

In a cyclone the central air pressure is lower than that of the surrounding environment, and the flow of circulation is clockwise in the southern hemisphere and counterclockwise in the northern hemisphere. Cyclones are also characterized by low-level convergence and ascending air within the system. An anticyclone's central air pressure is higher than that of its surroundings, and the airflow is counterclockwise in the southern hemisphere and clockwise in the northern hemisphere. Anticyclones are usually characterized by low-level divergence and subsiding air.

Vertical air movements are associated with both cyclones and anticyclones. In cyclones, air close to the ground is forced inward towards the centre of the cyclone, where pressure is lowest. It then begins to rise upward, expanding and cooling in the process. This cooling increases the humidity of the rising air, which results in cloudiness and high humidity in the cyclone. In anticyclones, the situation is reversed. Air at the centre of an anticyclone is forced away from the high pressure that occurs there. That air is replaced in the centre by a downward draft of air from higher altitudes. As this air moves downward, it is compressed and warmed. This warming reduces the humidity of the descending air, which results in few clouds and low humidity in the anticyclone.

Distinctive weather patterns are associated with both cyclones and anticyclones. Cyclones (commonly known as lows) are generally indicators of rain, clouds and other forms of adverse weather. Anticyclones (commonly referred to as highs) are predictors of fair weather.

LOCAL WINDS

Winds generated due to local differences in temperature on account of topographic and orographic variations are called local winds. These are of local importance and hence influence over very small area. The vertical and horizontal speed of these winds is felt within a short distance of 1 to 100 km.

Sea and land breezes

In coastal regions, sea breezes and land breezes can be important factors in a location's prevailing winds. The sea is warmed by the sun more slowly because of water's greater specific heat compared to that of land. As the temperature of the surface of the land rises during day time, the land heats the air above it by conduction. The warm air is less dense than the surrounding environment and so it rises. The cooler air above the sea, now with higher sea level pressure, flows inland into the lower pressure, creating a cooler breeze near the coast. The strength of the sea breeze is directly proportional to the temperature difference between the land mass and the sea. The velocity normally varies from 12 to 20 km/hour. This is more pronounced in summer than in winter. It starts at around 10 a.m., reaches its maximum at about 2 p.m. and declines thereafter till 8 p.m. to be replaced by land breeze. Sea breeze modulates diurnal temperature variation and the range between day and night temperature decreases.

At night, the land cools off more quickly than the ocean. This temperature change causes the daytime sea breeze to dissipate. When the temperature onshore cools below the temperature offshore, the pressure over the water will be lower than that of the land, establishing a land breeze.

Mountain and valley breezes

An interchange of air between the mountain slope and valley due to unequal heating and cooling of the two places is known as mountain and valley breeze. During daytime, the mountain slopes become more heated; the air over it expands leading to decrease in air density relative to the valley. Thus, a pressure gradient is established from valley to the mountain slope. The wind blows from valley to the mountain. This rising air sliding up the mountain slope is known as valley breeze. Its speed reaches a maximum at about 4 p.m. The valley breeze is also known as upslope wind or anabatic wind.

At night, there is a reverse process as denser cold air at higher elevation drains into depressions and valleys. Down slope winds are caused by greater air density on the slope than at some distance, horizontally from it. The wind is associated with surface cooling of the slope. If the air drains down slope into the valley, a mountain wind develops simultaneously along the axis of valley. This flows towards the plain, where it replaces warmer less dense air. The maximum velocity occurs just before sunrise at the time of maximum diurnal cooling.

Valley breezes are more common in summer and mountain breezes in winter. Valley breeze is more intense and stronger than the mountain breeze. Valley breezes may cause cumulus clouds over the mountain peaks and afternoon thunderstones. Mountain breezes cause radiation fogs in the early morning over the valley in winter. These breezes cause orographic rains and influence the local climate.

Winds due to topographic barriers

Over elevated surfaces, heating of the ground exceeds the heating of the surrounding air at the same altitude. It creates a thermal low over the terrain and changes the wind circulation of the region. Hills and valleys substantially distort the airflow by increasing friction between the atmosphere and landmass by acting as a physical block to the flow, deflecting the wind parallel to the range just upstream of the topography, which is known as a barrier jet. This barrier jet can increase the low level wind by 45%. Wind direction also changes because of the contour of the land. If there is a pass in the mountain range, winds will rush through it with considerable speed. The airflow can remain turbulent and erratic for some distance downwind into the flatter countryside. Cool winds accelerating through mountain gaps have been given regional names. In mountainous areas, local distortion of the airflow becomes severe. Jagged terrains combine to produce unpredictable flow patterns and turbulence. Strong updrafts, downdrafts and eddies develop as the air flows over hills and down valleys. In mountainous parts of the world subjected to relatively consistent winds, a moister climate usually prevails on the windward side of a mountain than on the leeward or downwind side.

Winds that flow over mountains down into lower elevations are known as downslope winds. These winds are warm and dry. In Europe, downwind of the Alps is known as *Foehn*. In Argentina, the local name for downsloped winds is *Zonda*. In Java, the local name for such winds is *Koembang*. In New Zealand, they are known as the *Nor'west arch*. In the Great Plains of the United States, the winds are known as a *Chinook*.

EFFECT OF WIND ON CROP PRODUCTION

Wind direction and velocity have significant influence on crop growth. Wind influences crop production by both physiological and mechanical ways. Physiological impacts include increased transpiration, drying of the plants by replacing humid air by dry air in the intercellular spaces, increased turbulence in the atmosphere and availability of CO_2 and thereby increased photosynthesis. But beyond a certain wind speed the rate of photosynthesis becomes constant. Mechanical impacts of wind include damages to the plant shoots, lodging of plants (paddy, wheat, maize, sugarcane, banana, *etc.*), flower and fruit shedding, uprooting of shallow rooted crops and trees, chilling injuries, soil erosion and poor aeration in root zone by soil deposition.

Beneficial impact of wind

- Wind increases the turbulence in atmosphere, thus increasing the supply of carbon dioxide to the plants resulting in greater photosynthesis rates
- Moderate wind helps in dispersal of pollens, thus increasing pollination in certain agricultural crops
- Wind helps in dispersal of seeds in some trees and shrubs which promotes natural propagation
- Wind prevents frost by disrupting a temperature inversion
- Calm to moderate wind favours dew deposition which benefits the crops under moisture stress in arid areas
- A wind speed of more than 15 km/hour favours winnowing operations

Harmful effects of wind

- Leaves become smaller and thicker, thus reducing photosynthetic area
- Direct mechanical effects are the breaking of plant structures, lodging of cereal crops, or shattering of seeds from panicles

- When wind is hot, desiccation of the plants takes place with high transpiration rate
- The hot and dry wind makes the cells expanding and induces early maturity; it results in the dwarfing of plants
- Under the influence of strong wind the shoots are pressurised and get deformed
- Strong winds cause uprooting of shallow rooted crops like papaya and banana
- In the coastal areas, winds carrying salt sprays can have harmful effect on susceptible crops
- Winds stronger than 10 km/hour can lift sand in arid regions leading to formation of sand dunes
- Strong winds develop xerophytic characters with increase in root to shoot ratio of plants
- Heavy wind during flowering reduces pollination, causes flower shed, increases sterility and reduces fruit set in all crops
- Soil and sand particles blown by wind strike the leaves of banana, sugarcane, maize, *etc.* making punctures, abrasions, lesions and tear the leaves into pieces and strips

Management of crops under high winds

- Growing of trees and tall crops across the direction of prevailing wind to reduce the physiological and mechanical damage to crops
- Adjusting the direction of planting
- Providing mechanical support to the tall crops like banana
- The effects of wind on evaporation can be avoided by using proper method of irrigation
- The damaging effect of wind can be reduced over a limited area by the use of shelterbelts (rows of trees planted for wind protection) and windbreaks (any structure that reduce the wind speed). Windbreaks reduce the wind velocity and create favourable microclimate
- Crops may be protected with preventive measures by weather forecasts

Chapter 5

Solar Radiation

Sun is the prime source of energy injected into the earth's atmosphere. The sun is the nearest star to the earth. It is a hot ball of glowing gases at the heart of the solar system. Its radiant energy is practically the only energy source to the earth. Very small and insignificant quantities of energy are available from other sources such as the interior of the earth, the moon, and other stars. Without the sun's intense energy and heat, there would be no life on earth.

NATURE AND PROPERTIES OF SOLAR RADIATION

The mean distance of the sun from the earth, also known as one astronomical unit (1 AU), is approximately 1.496×10^8 km, though the distance varies as the earth revolves round the sun in an elliptical orbit. At this average distance, light travels from the sun to earth in about 8 minutes and 19 seconds. The energy of the sunlight supports almost all life on earth by photosynthesis, and drives earth's climate and weather. The minimum sun-earth distance is about 0.983 AU and the maximum is approximately 1.017 AU. The earth is at its closest point to the sun (perihelion) on around January 3 and at its farthest point (aphelion) on around July 4.

The sun makes up about 99.86% of the mass of the entire solar system. The visible disk or photosphere has a radius of 6.599×10^5 km, and the solar mass is 1.989×10^{30} kg. The sun is a completely gaseous body. The chemical composition (by mass) of the outer layers of sun is 71% hydrogen, 26.5% helium, and 2.5% heavier elements like oxygen, carbon, neon, and iron among others. Since the sun is not a solid body, different parts of the sun rotate at different rates. At the equator, the sun spins once about every 25 earth days, but at its poles the sun rotates once on its axis every 36 days.

The physical structure of the sun is complex. The sun formed about 4.567 billion years ago from the gravitational collapse of a region within a large molecular cloud. Most of the matter gathered in the centre, while the rest flattened into an orbiting disk that would become the Solar System. The innermost region, the 'core', is the densest and hottest part of the sun. It is composed of highly compressed gases at a density of 100–150 g/cm^3.

Outside the core is the 'interior' which contains practically all of the sun's mass. The core and interior are thought to be a huge nuclear reactor. It generates energy by nuclear fusion of hydrogen nuclei into helium. In its core, the sun fuses about 620 million metric tons of hydrogen each second. These thermonuclear reactions supply the energy radiated by the sun. The energy is first transferred to the surface of the sun and then radiated into space. The surface of the sun, called the 'photosphere', is the source of most of the visible radiation arriving at the earth's surface. It is composed of very low density gases. The photosphere is the source of radiation flux to space because it has the capability to emit and absorb a continuous spectrum of radiation. Outside the photosphere is the solar atmosphere, called the 'reversing layer' which is several hundred kilometres deep and almost transparent. This layer contains vapours of almost all of the known elements found on the earth. Outside the reversing layer is the 'chromosphere', which is about 25,000 km deep. In this zone, the short-lived, solar flares occur in the clouds of hydrogen and helium. These brilliant flares are a source of intense bursts of ultraviolet and radio wave radiation. The solar flares also eject streams of electrically charged particles called corpuscles, which, on reaching the earth's surface, disturb its magnetic field. The temperature in the chromosphere is several times higher than that of the photosphere. The outermost portion of the sun is called the 'corona'. It is composed of extremely rarefied gases known as the solar winds. These winds consist of very sparse ions and electrons moving at very high speeds and are thought to extend into the solar system. The solar radiation flux is the composite result of several layers of the sun.

Insolation is the total amount of incoming solar radiation received on a given horizontal surface area during a given time. It is also called solar irradiation and expressed as 'hourly irradiation' if recorded during an hour or 'daily irradiation' if recorded during a day. Radiant flux density or the intensity of radiation is the amount of radiant energy received on a unit area of surface, which is held perpendicular to the solar beam in unit time. The unit for insolation recommended by the World Meteorological Organisation is the megajoule per square metre (MJ/m^2) or joule per square millimetre (J/mm^2). Alternate units used are the 'langley' (1 thermochemical calorie per square centimetre or 41,840 J/m^2) and watt-hour per square metre (Wh/m^2). If this energy is divided by the recording time in hours, it is then a density of power called irradiance, expressed in watts per square metre (W/m^2).

The radiative heat transfer process is independent of the presence of matter. It can move heat even through empty space. All bodies emit radiation and the wavelength (or frequency) and energy characteristics (or spectrum) of that radiation are determined solely by the body's temperature. Radiation goes through a transformation when it encounters other objects (solid, gas or liquid). That transformation depends on the physical properties of that object and it is through this transformation that radiation can transfer heat from the emitting body to the other objects.

Emission, absorption, reflection and transmission

A part of the incident radiation on the surface is absorbed while a part is reflected and the remaining is transmitted. This can be put in the form of an equation.

$$\alpha + r + t = 1$$

where, α is the aborptivity; r is the reflectivity and t is the transmissivity.

Emissivity: It is the ratio of the emitted radiation of a given surface at a specified wavelength to the emittance of an ideal blackbody at the same wavelength and temperature. For other than a blackbody the value of emissivity is always less than one and for blackbody the emittance value is one.

Absorptivity: For an object this is the ratio of the radiant energy absorbed to the total amount incident upon the same object. Like emissivity, the absorptivity of a blackbody is one and the values are less than one for other than blackbodies. Natural bodies like the sun and the earth are nearly perfect blackbodies.

Reflectivity: It is defined as the ratio of the radiant energy reflected to the total incident radiation upon that surface. It is expressed in per cent.

Transmissivity: This is the ratio of transmitted radiation to the total incident radiation upon that surface.

The ultimate source of energy for all atmospheric process is solar energy. This energy travels through space at a speed of light as electromagnetic radiation and transmitted to the earth as a sequence of waves. The amplitude or height of the electromagnetic waves is related to the amount of energy transfer, whereas the type or quality of energy is described by its wavelength. In fact, all matters in the universe having temperature above absolute zero (the temperature at which all atomic or molecular motion stops) emit radiation. The intensity and wavelength of emission from a body is related to its temperature as described by Stefan-Boltzmann's law and Wien's displacement law, respectively.

LAWS OF RADIATION

Radiation of heat from sun to earth's surface through space and atmosphere reaches in the form of electromagnetic waves with the same speed as light. The wavelength of electromagnetic radiation is given by the equation

$$\lambda = c/v$$

where λ is the wavelength, the shortest distance between consecutive crests in the wave trans; c is the constant equal to the velocity of light, 3×10^{10} cm/s; and v is the frequency, number of vibrations or cycles per second. The radiation reaching the earth surface from the sun through the space and atmosphere, and radiating back from earth to atmosphere follows certain physical laws known as radiation laws, as discussed below.

Planck's law

The electromagnetic radiation consists of a stream or a flow of particles or quanta. Each quantum having an energy content E determined by

$$E = hv$$

where, h is the Planck's constant (having a value of 6.625×10^{-34} Js) and v is the frequency. The greater the frequency (*i.e.*, shorter the wavelength), the greater is the energy content of the quantum.

Stefan-Boltzmann's law

The intensity of radiation emitted by a radiating body (E) is directly proportional to the fourth power of its absolute temperature.

$$E = \sigma T^4$$

where, E is the emissivity of the body; σ is the Stefan-Boltzmann constant; and T is the absolute temperature.

$$\sigma = 5.67 \times 10^{-8}\ \text{W/m}^2/\text{K}^4$$

Wien's displacement law

The wavelength of maximum emission (λ_{max}) is inversely proportional to the absolute temperature of the radiating body.

$$\lambda_{max} = \omega / T$$

where, λ_{max} is expressed in micrometre (µm), T is the absolute temperature expressed in K and ω is the Wien's constant having value 2897 µmK.

Kirchoff's law

The law states that the absorptivity ε of an object for radiation of a specific wavelength (λ) is equal to its emissivity α for the same wavelength. The law is expressed as

$$\varepsilon (\lambda) = \alpha (\lambda)$$

FACTORS AFFECTING SOLAR RADIATION

Variations in radiation received over different regions of the earth make difference in the weather and climate on earth's surface. The amount of solar radiation reaching the earth's surface depends on two factors, the amount of radiation available at the top of the atmosphere and the reduction of that amount due to atmospheric attenuation. The amount of energy received at the top of the atmosphere is affected by four factors; solar output, solar constant, the altitude of the sun and the length of the day.

Solar output

Solar energy originates from nuclear reactions within the sun's hot core and is transmitted to its surface by radiation and hydrogen convection. However, most energy produced in the interior is confined by its hydrogen convection zone and does not reach the outer surface. The surface of the sun (photosphere) has an average temperature of about 5800 Kelvin from which solar radiation is emitted to the space. The Stefan-Boltzmann's law mathematically expresses the rate of radiation of a blackbody. Assuming the sun as a blackbody, at a temperature of 5800 K, the rate of radiation from sun is calculated to be about 64 million watts/m^2. For the sun's surface area of 6.03×10^{18} m^2 the solar energy output would be of the order of 3.84×10^{26} watt. Solar radiation may fluctuate slightly due to periodic sunspot activities. The wavelength of peak intensity of emission of a blackbody is inversely proportional to the absolute temperature of the body and can be estimated by Wien's displacement law. For solar radiation at 5800 K, the peak radiation is attained at about 0.5 µm and the wavelength varies mainly between 0.2 and 4.0 µm. The solar radiation is shortwave in nature and consists of about 7% is ultraviolet (0.2–0.4 µm), 41% visible light (0.4–0.7 µm) and 52% near infrared (> 0.7 µm).

Solar constant

The electromagnetic energy from the sun moving through space is not depleted as it moves towards the earth. Radiation travelling through space carries the same amount of energy and has the same wavelength as when it left the solar surface. However, as the distance from the sun increases, it is distributed over greater area and thus the intensity of radiation diminishes in proportion to the distance squared. This relationship is known as inverse square law. By dividing the total solar emission (3.84×10^{26} W) by the area of the imaginary sphere surrounding the sun ($4\pi R^2$, where R is the earth's average distance from the sun *i.e.*, 1.5×10^{11} m or 150 million km), the amount of solar energy received by a surface perpendicular to the incoming rays at the mean earth-sun distance can be determined. This incoming solar radiation intercepted by the earth at the top of the atmosphere directly facing the sun is estimated to be about 1368 watts per square metre. This amount of energy is known as the total solar irradiance or the solar constant. The solar constant is only approximately constant. Depending on the distance of the earth from the sun, its value ranges from approximately 1360 to 1380 W/m^2. Of this energy, approximately 31% is scattered back to space, 43% is absorbed by the earth's surface, and the atmosphere absorbs 26%.

The actual solar energy intercepted by the earth is limited by the basic geometry of earth in its orbit. At any time, only half the earth is illuminated by the sun. Besides, the solar irradiance is maximum at a surface where incoming solar light comes at right angles *i.e.*, perpendicular to the ground surface. Because the earth is a sphere, only areas near the equator at midday are perpendicular to the path of incoming light. Everywhere else, the light comes obliquely. Thus solar irradiance decreases with increasing latitude. Averaged over the entire planet, the amount of sunlight arriving at the top of the earth's atmosphere is only one-fourth (The surface area of the earth is $4\pi R^2$. But at any time the solar light is intercepted by a surface equal to the cross sectional area of this sphere *i.e.*, πR^2.) of the total solar irradiance, or approximately 342 watts per square metre. There are significant seasonal changes in the receipt of incoming solar radiation on the earth though the sun emits a nearly constant amount of radiation. As the earth revolves around the sun in an elliptical path, the distance between the earth and the sun varies over the course of a year. The earth-sun distance is minimum (at a point on its orbit called perihelion, 147.3 million km) on third January and maximum (at aphelion, 152.1 million km) on fourth July.

Altitude of the sun

The altitude of the sun (the angle between its rays and a tangent to the earth's surface at the point of observation) determines the angle of incidence of the solar radiation. When the angle of incidence is high, the beam spreading decreases *i.e.*, radiation becomes more concentrated per unit area. Beam spreading is the increase in surface area over which radiation is distributed in response to a decrease of solar angle. The sun's altitude is determined by the latitude of the site, time of the day and the season. Axis of rotation of the earth is tilted all year 23.5° from a line perpendicular to the plane of the earth's orbit around the sun. As a result, the sun is directly overhead at the Tropic of Cancer (23.5° N latitude) on 21st June in the northern hemisphere, making it the longest day in the northern hemisphere and the shortest day in the southern hemisphere. This day is called as summer or winter solstice in the northern or southern hemisphere, respectively. Similarly, the sun is directly overhead at the Tropic of Capricorn (23.5° S latitude) on 22nd December in the southern hemisphere, making it the longest day in the southern hemisphere and the shortest day in the northern hemisphere. This day is called as summer or winter solstice in the southern or northern hemisphere, respectively.

On March 21 (spring or vernal equinox) and September 23 (autumnal equinox), the sun is directly over the equator. The strength of sunlight increases from sunrise until noon, when the sun is directly overhead along the equator. After noon, the strength of sunlight decreases until the sun sets at 6.00 p.m. During the equinoxes, the tropics (from 23.5° N to 23.5° S latitude) receive about 90% of the energy that falls on the equator, the mid-latitudes (45°) receive roughly 70%, and at the Arctic and Antarctic Circles (66.5°) about 40% is received. The change in noontime solar angle over the course of a year can cause significant differences in the intensity of sunlight hitting the surface due to beam spreading. Beam spreading of the noontime sun is greater in winter than summer and the difference increases with latitude. When sunlight approaches the surface at low angles, the beam of sunlight also has to pass through a greater amount of atmosphere and more energy is lost due to atmospheric effects.

Period of daylight

The tilt of the earth's axis also affects the length of day and night and thus influences the energy receipt on the earth. The longer the time the sun shines, the greater is the quantity of radiation received. At the equator, the day and night are both close to 12 hours long throughout the year. The day length within Arctic and Antarctic circles (beyond 66.5° N and S latitudes) varies between 0 and 24 hours from winter to summer. During June solstices, in northern hemisphere the day length gradually increases from 12 hours at equator to 24 hours at and beyond 66.5° N latitude and in southern hemisphere the length of nights increases in a similar way. During December solstice, the pattern is reversed.

The amount of solar energy received at the top of the atmosphere due to combination of all these factors are given in table 5.1. The polar regions receive their maximum amounts of solar radiation during their summer solstices, which is the period of continuous days. The amount received during December solstice in the southern hemisphere is theoretically greater than the amount received by the northern hemisphere during June solstice. The equator has two radiation maxima at the equinoxes and two minima at the solstices.

Table 5.1 Daily solar radiations on a horizontal surface outside the atmosphere in W/m²

Date	90° N	70	50	30	0	30	50	70	90° S
December 22	0	0	86	227	410	507	514	526	559
March 21	0	149	280	378	436	378	280	149	0
June 21	524	492	482	474	384	213	80	0	0
September 23	0	147	276	373	430	372	276	147	0

(*Source*: Berger, 1996)

DEPLETION OF SOLAR RADIATION

The loss of solar energy in passing through the atmospheric layers is called the atmospheric deflection. The longer the path traversed, the greater the amount of radiant energy depleted. If the sun's radiation was not filtered or depleted in some manner, our planet would be soon so hot for life to exist. This depletion of solar radiation is accomplished through dispersion, scattering, reflection, and absorption.

Dispersion

Since the earth's axis is inclined at an angle of 23.5°, the sun's rays are received on the surface of earth at varying levels of incidence, depending on the position of earth. When the sun's rays are not perpendicular to the surface of earth, the energy becomes dispersed or spread out over a greater area. If the available energy reaching the atmosphere is constant and is dispersed over a greater area, the amount of energy at any given point within the area decreases, and therefore the temperature is lower. Dispersion of insolation in the atmosphere is caused by the rotation of earth. Dispersion of insolation also takes place with the seasons in all latitudes, but especially in the latitudes of the polar areas.

Scattering

About 25% of the incoming solar radiation is scattered or diffused by the atmosphere. Scattering is a phenomenon that occurs when solar radiation passes through the air and some of the wavelengths are deflected in all directions by molecules of gases, suspended particles, and water vapour. These suspended particles then act like a prism and produces a variety of colours. Various wavelengths and particle sizes result in complex scattering effects that produce the blue sky. Scattering is also responsible for the red sun at sunset, varying cloud colours at sunrise and sunset, and a variety of optical phenomena. Scattering always occurs in the atmosphere, but does not always produce dramatic settings. Under certain radiation wavelength and particle size conditions all that can be seen are white clouds and a whitish haze. This occurs when there is a high moisture content (large particle size) in the air and is called diffuse reflection. About two-thirds of the normally scattered radiation reaches earth as diffuse sky radiation. Diffuse sky radiation may account for almost 100% of the radiation received by polar regions during winter.

Reflection

Reflection is the process whereby a surface turns a portion of the incident back into the medium through which the radiation came. Thus, a certain amount of energy is lost and it does not play any role in heating the atmosphere. The reflectivity of a substance, usually expressed as a percentage of the incident radiation reflected, is known as its albedo or reflection coefficient. It differs from place to place as well as from time to time in the same locality. It depends upon the amount of cloudiness and atmospheric impurities, the obliquity of the sun's rays, and the nature of the surface. The albedo for the earth as a whole, also known as the planetary albedo, is 35%.

Absorption

Absorption has been defined as the process in which the incident radiation is retained by a substance and is irreversibly converted to some other form of energy. There is a certain amount of radiant energy loss by the process of absorption in the atmosphere. Earth and its atmosphere absorb about 64% of the insolation. Land and water surfaces of earth absorb 51% of this insolation. The remaining 13% is directly absorbed by ozone, carbon dioxide, and water vapour. The gases are selective absorbers and absorb the insolation at certain wavelengths. For example, ozone absorbs only a small percentage of the insolation. The portion or type the ozone does absorb is critical since it reduces ultraviolet radiation to a level where animal life can safely exist. Nitrogen is a very poor absorber of insolation. The most important absorption occurs with carbon dioxide and water vapour which absorb strongly over a broader wavelength band. Clouds are by far the most important absorbers of radiation at essentially all wavelengths. In sunlight, clouds reflect a high percentage of the incident solar radiation and account for most of the brightness of earth as seen from space. There are regions, such as areas of clear skies, where carbon dioxide and water vapour are at a minimum and so is absorption. These areas are called atmospheric windows and allow insolation to pass through the atmosphere relatively unimpeded.

INCOMING SHORTWAVE RADIATION

When the sun's energy reaches the earth it is partially absorbed in different parts of the climate system. The absorbed energy is converted back to heat, which causes the earth to warm up and makes it habitable. Absorption of solar radiation is uneven in both space and time giving rise to the intricate pattern and seasonal variation of the climate on the earth.

Solar radiation occurs over a wide range of wavelengths. However, the energy of solar radiation is not divided evenly over all wavelengths but, is rather sharply centred on the wavelength band of 0.2–2.0 micrometres (μm = one millionth of a metre). The main range of solar radiation includes ultraviolet radiation (UV, 0.001–0.4 μm), visible radiation (light, 0.4–0.7 μm), and infrared radiation (IR, 0.7–100 μm). The radiations under visible region of wavelengths (0.4–0.7 μm) are also called as photosynthetically active radiations (PAR). These are subdivided into various wave bands based on subjective colours experienced by humans. The sun is assumed to be a blackbody, whose surface temperature is approximately 6000 K. Hence, by Wien's law the wavelength of maximum emission is 0.483 μm. Since this wavelength comes under shortwave band, the solar radiations are known as shortwave radiations.

Table 5.2 Energy content of electromagnetic spectrum in various visible colour bands

Colour	Wavelength band (μm)	Irradiance (W/m^2)	% of total irradiance
Violet	0.390–0.455	108.85	7.96
Blue	0.455–0.492	73.63	5.39
Green	0.492–0.577	160.00	11.70
Yellow	0.577–0.597	35.97	2.63
Orange	0.597–0.622	43.14	3.16
Red	0.622–0.770	212.82	15.57

(*Source*: Mavi and Tupper, 2004)

As the sun's energy spreads through space that contains almost no interfering matter its spectral characteristics do not change. However, the energy flux drops monotonically as the square of the distance from the sun. Thus, when the radiation reaches the outer limit of the earth's atmosphere several hundred kilometres over the earth's surface, the radiative flux is approximately 1360 W/m^2. Solar radiation reaching the top of the atmosphere has to traverse through the layers of atmosphere before reaching the earth's surface. During its passage, the radiation is attenuated by a variety of processes and the combined process is known as atmospheric depletion. When radiation strikes an object, it may be absorbed, transmitted or redirected. The amount of energy absorbed by an object depends on the intensity of radiation and the object's absorptivity. About 20% of radiation entering the atmosphere may be absorbed at certain wavelengths by atmospheric gases such as ozone and water vapour. Absorption of radiant energy causes an increase in temperature. Substances such as water and air, which are transparent to certain wavelengths of radiation, may simply transmit energy. The atmosphere is quite transparent to incoming shortwave solar radiation. On average, 49% of the energy reaching the top of atmosphere is transmitted though the atmosphere and received at the earth's surface. Radiation that is transmitted does not contribute energy to the object. Some radiation may bounce off the object without being absorbed or transmitted. Reflection and scattering are responsible for redirecting incoming solar radiation. About 22% is reflected back to space by the atmosphere and clouds whereas 9% is reflected from earth's surface. The reflectivity of a substance, usually expressed as a percentage of the incident radiation reflected, is known as its **albedo**. The albedo varies with the colour and composition of the earth's surface, the season, and the angle of the sun's rays. The albedo values of some selected surfaces on earth are given in the table 5.3. However, the albedo of the earth as a whole is known as planetary albedo and is about 31%.

Table 5.3 Albedo of surfaces on earth (shortwave radiation)

Surface	Albedo (%)
Fine sandy soil	37
Dark black soil	14
Moist black soil	8
Deciduous forest	17
Pine forest	14
Desert scrubland	20–29
Ice sheet with water	26
Sea ice	36
Dense clean dry snow	86–95
Water surface at 30° latitude	6–9

(*Source*: Barrett, 1992)

The direction of a photon of radiation is changed by interaction with atmospheric gases and aerosols through reflection or scattering. Reflection is a process whereby radiant light bounces back from an object at the same angle and intensity at which it encounters a surface. Scattering produces a large number of weaker rays travelling in different directions. Whether solar radiation is reflected or scattered depends largely on the size of intervening particles and the wavelength of the radiation. Three types of scatterings are distinguished. For gas molecules smaller (about one-tenth) than the radiation wavelength, scattering occurs in all directions and is inversely proportional to wavelength. This type of scattering is called Rayleigh scattering. Scattering of blue light is 10 times more, creating the day time blue sky. When water droplets or aerosol particles, with similar size (0.1- 0.5μm) to the radiation wavelength are present, most of the light is scattered forward. This is Mie scattering giving the greyish appearance to polluted atmosphere. The water droplets in clouds behave like lenses and scatter all wavelengths of incoming radiation almost equally. This is called non-selective scattering. Clouds are by far the most important agents in non-selective scattering and exert a tremendous impact on the global receipt of solar radiation by reflecting large amount of energy back to space. Radiation undergoes multiple scattering within a cloud or between low clouds and a snow covered surface. Atmospheric scattering gives rise to diffuse radiation. It is measured separately from the direct beam radiation. For average cloudiness, the diffuse radiation is about 50% of the total radiation.

Heat energy flows from a region of higher temperature to one of lower temperature till equilibrium in temperature is established. This flow of energy can occur in three ways; conduction, convection and radiation. All these mechanisms of heat transfer can operate simultaneously. These processes are responsible for the transfer of heat between the sun and earth and between earth's surface, the atmosphere and outer surface.

Radiation is the transfer of energy (both heat and light) by electromagnetic waves which do not require a medium such as air for their transmission, at a speed of light. It is the only heat transfer mechanism by which solar energy reaches our planet. During emission, a body releases electromagnetic energy at different wavelengths. The intensity of emission from a body varies with wavelength, temperature and efficiency of emission. Bodies that emit radiation with perfect efficiency are termed as blackbodies. The sun as well as the earth is close to it. When solar radiation travels through the vacuum of outer space, there is no modification of its intensity, direction or wavelength. However, when it enters the atmosphere, only some of the radiation can pass unobstructed to the surface. A clear dry atmosphere might transmit as much as 80% of the incoming radiation as direct beam radiation. In contrast, when it is cloudy or hazy, only a small fraction will reach the surface as direct radiation and others will either be reflected back to space or reach the surface as diffused radiation. Conduction is the transfer of energy in a medium from one molecule to the next in the presence of a temperature gradient. The medium as a whole experiences no molecular movement. As air is a poor conductor, this type of heat transfer is negligible in the atmosphere. But it is important in the ground surface that receives more radiant energy than the atmosphere and is hotter. Conduction affects air temperature by transferring energy between the soil surface and the bottom molecular layer of the air. Convection is the transfer of energy in fluids (including gases) that are able to circulate internally and distribute heated parts of mass. Here energy is transferred as the molecules themselves move. Convection is the chief means of atmospheric heat transfer due to the low viscosity of air and its almost continual motion.

Table 5.4 Disposal of shortwave solar radiation

Solar energy	W/m^2
Incident on the top of the atmosphere (a+b+c+d)	342
Reflected by clouds, aerosols, and atmosphere (a)	77
Reflected from the earth (b)	30
Total reflected (a+b)	107
Absorbed by the atmosphere (c)	67
Absorbed by the earth (d)	168
Total absorbed by earth-atmosphere system (c+d)	235

(*Source*: Kiehl and Trenberth, 1997)

OUTGOING LONGWAVE RADIATION

Like sun, earth also acts as a blackbody and radiates energy as electromagnetic waves. The earth's surface reflects (*i.e.*, returns the radiation back to space in more or less the same spectrum) part of the energy it receives from the sun. The earth behaves as a blackbody with mean temperature of about 300 K. Thus, by Wien's displacement law the wavelength of maximum emission is calculated to be 9.66 μm. Since this wavelength comes under long-wave band, the solar radiations are known as longwave radiations. About 99% of this radiation is emitted in the infrared range from 4 to 120 μm, with a peak near 10 μm. The atmosphere has a varying range of temperature from cold to terrestrial temperatures up to 323 K. So it also radiates to earth in the longwave band.

This outgoing radiation from the earth is called terrestrial radiation and should be equal in amount to the incoming radiation. A significant part of the earth's reflectivity can be attributed to clouds. In climatology, the reflectivity of a planet is referred to as the albedo and is expressed as a fraction. The albedo of earth depends on the geographical location, surface properties, and the weather. On the average, however, the earth's albedo is about 0.31. This fraction of incoming radiation is reflected back into space. The other 0.69 fraction of the incoming solar radiation is absorbed by the earth.

The earth's atmosphere is transparent to most of the incoming shortwave solar radiation but it selectively absorbs a portion of the outgoing longwave or infrared radiation. This is due to the net infrared absorption capacity of water vapour, carbon dioxide and other trace gases. Water vapour is a strong absorber at several wavelength bands of 0.7 to 8.0 μm and above 13 μm and it accounts for approximately 89% of the total infrared absorption. Carbon dioxide is the second most important and abundant gas, accounting for about 7.5% of the absorption. It absorbs strongly at 2.7 and 4.3 μm and above 13 μm. Ozone is a perfect absorber of ultraviolet radiation upto wavelength band of 0.3 μm. Another peak due to absorption by ozone is in wavelengths of 9.4 to 9.8 μm. The clouds absorb in all wavelengths. A large part of the radiation absorbed by the atmosphere is sent back to the earth's surface as counter radiation. This counter radiation prevents the earth's surface from excessive cooling at night.

Longwave radiations escape to space between 8.5 and 11.0 μm, known as the atmospheric window. In the atmospheric window, gases are relatively transparent to the outgoing thermal infrared radiation. Out of the 390 W/m^2 terrestrial radiation, 40 watt/m^2 is escaped into space whereas 350 W/m^2 are absorbed in the atmosphere. The opaqueness of the atmosphere to infrared radiation relative to its transparency to shortwave radiation is commonly referred as the greenhouse effect and the gases such as water vapour, carbon dioxide, *etc.* that absorb the infrared radiation are called greenhouse gases.

Other greenhouse gases are methane, nitrous oxide and ozone. When greenhouse gas molecules absorb thermal infrared energy their temperature rises. Greenhouse gases then radiate an increased amount of thermal infrared energy in all directions. Some of it spreads downward and ultimately comes back into contact with the earth's surface, where it is absorbed. Out of the 350 W/m^2 absorbed by the atmosphere, 324 W/m^2 energy again reradiates by the atmosphere to the ground surface. The temperature of the surface becomes warmer than it would be if it were heated only by direct solar radiation. Some of the heat escapes directly to space, and the rest is transferred to higher and higher levels of the atmosphere, until the energy leaving the top of the atmosphere matches the amount of incoming solar energy. Because the maximum possible amount of incoming sunlight is fixed by the solar constant the natural greenhouse effect does not cause a runaway increase in surface temperature on earth. The quantitative disposal of longwave radiation from the earth-atmosphere system is summarised in table 5.5.

Table 5.5 Disposal of longwave radiation

Longwave radiation	Watts/m^2
Emitted by the earth's surface (a+b)	390
Lost to space (a)	40
Absorbed by the atmosphere (b)	350
Emitted by the atmosphere and clouds (c+d)	519
Lost to space from atmosphere (c)	195
Back radiation from atmosphere absorbed by earth (d)	324
Total outgoing longwave radiation (a +c)	235

(*Source*: Kiehl and Trenberth, 1997)

THERMAL RADIATION

Thermal radiation is a process by which energy, in the form of electromagnetic radiation, is emitted by a heated surface in all directions and travels directly to its point of absorption at the speed of light. Thermal radiation does not require an intervening medium to carry it. Thermal radiation ranges in wavelength from the longest infrared rays through the visible-light spectrum to the shortest ultraviolet rays. The intensity and distribution of radiant energy within this range is governed by the temperature of the emitting surface. All matter with a temperature greater than absolute zero emits thermal radiation and the rate of thermal radiation emission increases with increasing temperature. The total radiant heat energy emitted by a surface is proportional to the fourth power of its absolute temperature (the Stefan–Boltzmann law). The rate at which a body radiates (or absorbs) thermal radiation depends upon the nature of the surface as well. Objects that are good emitters are also good absorbers (Kirchhoff's radiation law). A blackened surface is an excellent emitter as well as an excellent absorber. If the same surface is silvered, it becomes a poor emitter and a poor absorber. A blackbody is one that absorbs all the radiant energy that falls on it. Such a perfect absorber would also be a perfect emitter.

Modes of heat transfer

The fundamental modes of heat energy transfer are conduction or diffusion, convection, radiation and advection.

Conduction: The process of energy transfer between objects that are in physical contact is called conduction or diffusion. In this mechanism, heat passes from point to point by means of transfer of adjacent molecules possessing higher kinetic energy to those possessing less energy. This type of heat transfer is negligible in atmosphere as air is a poor conductor but is important on ground.

Convection: The transfer of heat energy between an object and its environment, through movement of particles in fluids is called convection. The low viscosity of air and its consequent ease of motion make this the chief method of atmospheric heat transfer.

Radiation: Radiation is the transfer of energy to or from a body by means of the emission or absorption of electromagnetic radiation. This is the case with solar energy passing through space, where the earth's atmosphere only allows the passage of radiation at certain wavelengths and restricts at the others.

Advection: The transfer of energy from one location to another as a side effect of physically moving an object containing that energy is called advection.

ENERGY BUDGET

Weather and climate on earth are determined by the amount and distribution of incoming radiation from the sun. The earth system acts as a heat engine that moves heat from one part of the surface to another. It also moves heat from the earth's surface and lower atmosphere back to space. This flow of incoming and outgoing energy is earth's heat budget. For earth's temperature to be stable over long periods of time and for equilibrium climate, outgoing longwave radiation necessarily balances the incoming shortwave radiation. In other words, the energy budget at the top of the atmosphere must balance. This state of balance is called radiative equilibrium. However, there are great deal of fascinating atmosphere, ocean, and land phenomena that couple the two. Incoming radiant energy may be scattered and reflected by clouds and aerosols or absorbed in the atmosphere. The transmitted radiation is then either absorbed or reflected at the earth's surface. Radiant solar or shortwave energy is transformed into sensible heat, latent energy, potential energy, and kinetic energy before being emitted as longwave radiant energy. Energy may be stored for some time, transported in various forms, and converted among the different types, giving rise to a rich variety of weather or turbulent phenomena in the atmosphere and ocean.

About 31% of the solar energy that arrives at the top of the atmosphere is reflected back immediately to space by clouds, atmospheric particles or bright ground surfaces like sea ice and snow. This is the planetary albedo and this energy plays no role in earth's climate system. About 20% of incoming solar energy is absorbed in the atmosphere by water vapour, dust, ozone and other atmospheric gases. The remaining 49% pass through the atmosphere and are absorbed by the surface. Thus, the atmosphere and the surface of the earth together absorb 69 units of incoming solar radiation. So together, they must radiate that much energy back to space for the planet's average temperature to remain stable. However, the relative contribution of the atmosphere and the surface of earth to each process (absorbing sunlight versus radiating heat) are asymmetric. Most solar heating happens at the surface, while most radiative cooling happens in the atmosphere.

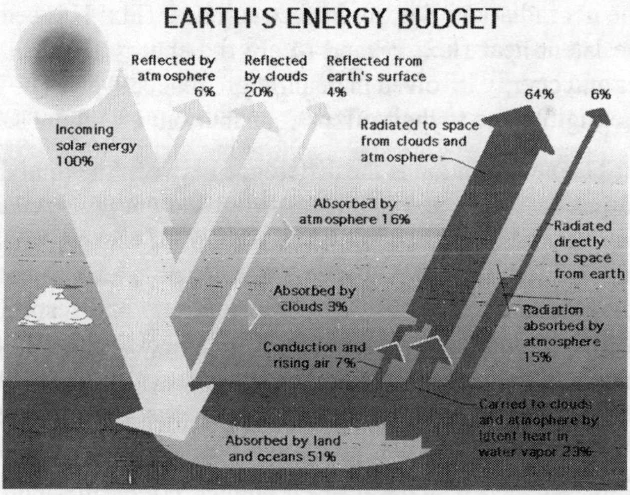

Fig 5.1 Earth's energy budget
(*Source*: https://ag.tennessee.edu/solar/Earth-Energy-Budget.aspx)

For the energy budget at earth's surface to balance, processes on the ground must get rid of this 49% of incoming solar energy that the ocean and land surfaces absorb. Energy leaves the surface through three processes; evaporation, convection, and emission of thermal infrared energy. About 23% of incoming solar energy leaves the surface through evaporation. Liquid water molecules absorb incoming solar energy, and they change phase from liquid to gas. The heat energy that it takes to evaporate the water is latent in the random motions of the water vapour molecules as they spread through the atmosphere. When the water vapour molecules condense back into rain, the latent heat is released to the surrounding atmosphere. An additional 7% of incoming solar energy leaves the surface through convection. Air in direct contact with the sun-warmed ground becomes warm and buoyant. In general, the atmosphere is warmer near the surface and colder at higher altitudes, and under these conditions, warm air rises, shuttling heat away from the surface. This is called the sensible heat energy. Finally, a net radiation of about 19% of incoming solar energy leaves the surface as thermal infrared energy radiated by atoms and molecules on the surface.

RADIATION BALANCE OVER A CROP SURFACE

The total of all incoming and outgoing radiations over a surface is zero. The radiation balance over a crop surface may be expressed by the following equation.

$$R_n = S + H + LE + PS + M$$

where, R_n is the net radiation; S is the surface soil heat flux; H is sensible heat flux; LE is the latent heat flux; PS and M are the energy fixed in plants by photosynthesis and energy involved in respiration, respectively. The PS and M are assumed negligible due to their minor contribution, *i.e.*, about 1-2% of R_n.

Net radiation: The net radiation is the difference between the total downward and upward radiation fluxes and is a measure of the energy available at the ground surface. Net radiation represents the amount of energy, which is used for various kinds of activities. It is dispensed as sensible heat, latent heat and also in physiological processes such as photosynthesis and respiration. The importance of this parameter is that it is the fundamental quantity of energy available at the earth's surface to drive the processes of evaporation, air and soil heat fluxes as well as other smaller energy consuming processes like photosynthesis, *etc.* Usually, the net radiation is positive between an hour after sunrise and an hour before sunset with a maximum at midday.

Soil heat flux: It is the flow of heat into or out of the soil through a process of conduction. Soil constitutes major storage for heat, acting as a sink for energy during the day and a source at night. Thus, the soil stores energy during the summer season and releases it to the air during winter season of the year.

Sensible heat flux: It is the flow of heat between the surface and air near the surface through a process of convection. This transfer of heat determines air temperature.

Latent heat flux: It is the flow of heat to and from the surface through evaporation of water or condensation. In other word, the amount of heat required in the process of evaporation and amount heat released in the process of condensation is called latent heat.

Photosynthesis: It is the quantum of radiation, which is utilised by the plant and converted into biomass.

Metabolic heat: It is the part of radiation which is utilised by the plant in respiration and in maintaining its canopy temperature.

RESPONSE OF CROP PLANTS TO RADIATION

The solar spectrum in the wavelengths between 0.39 and 0.77 μm is called light, which is visible. The visible radiation is very important in photosynthetic mechanism of plants. The solar radiation within this range of wavelengths *i.e.*, visible radiation is called photosynthetically active radiation (PAR). The carbon dioxide and water in the presence of light go through a process called the photosynthesis in the green leaves or other parts of the plant. In this process radiation is converted into biochemical energy and stored in chemical bonds *i.e.*, carbohydrates which deposits biomass or plant dry matter. Radiation is the major mode of energy exchange between plant and aerial environment. Solar energy provides the main energy input to plants. Most of this energy is converted into dry matter and drives other radiation exchange processes such as transpiration, respiration, *etc*. Quality, duration and intensity of light influence plant processes and development.

Quality of light

Spectral composition of solar radiation is the index of its quality. It consists of ultraviolet, visible, near-infrared and infrared wavelengths of which only radiation in the range of 0.3 to 1.0 μm are of importance in crop growth. The quality of radiation affects flowering, germination and elongation. Infrared radiation is associated with heat balance of the plant while light below 0.315 μm wavelength is detrimental. Quality of light influences plant development through its photosynthetic process. The solar spectrum of wavelength from 0.6 to 0.7 μm and from 0.4 to 0.5 μm are of strong photosynthetic and formative activity, while the wavelength from 0.5 to 0.6 μm is of weak photosynthetic and formative activity. Radiation between 0.3 and 0.4 μm wavelength produces formative effects leading to shorter plants with thick leaves. Mavi and Tupper (2004) divided the solar spectrum into eight broad bands on the basis of the physiological response of plants (Table 5.6).

Table 5.6 Physiological response of plants to different solar radiation wavelength bands

Wavelength band (μm)	Physiological response of plants
> 1.000	Most of this radiation absorbed by plants is transformed into heat without interfering with the biochemical processes.
1.000 to 0.700	Elongation effects on plants.
0.700 to 0.610	Very strong absorption by chlorophyll, the strongest photosynthetic activity, and in many cases strong photoperiodic activity.
0.610 to 0.510	Low photosynthetic effectiveness in the green segment and weak formative activity.
0.510 to 0.400	Strong chlorophyll absorption, strong photosynthetic activity, and strong formative effects.
0.400 to 0.315	Produces fluorescence in plants and a strong response by photographic emulsions.
0.315 to 0.280	Significant germicidal action.
< 0.280	Very strong germicidal action. When below 0.26 μm it can kill some plants. No such radiation reaches the earth's surface.

(Source: Mavi and Tupper, 2004)

Intensity of light

Light intensity measures its quantity or brightness. Intensity of light influences crop growth through its photosynthetic process. Production of photosynthates depends on compensation point and saturation light intensity. Compensation point is the maximum light intensity at which respiration rate equals photosynthesis rate. Higher the temperature, higher will be the compensation point. Saturation light intensity is the maximum intensity at which the rate of photosynthesis reaches a maximum value. Most crop plants reach light saturation at 1/5th of full sunlight. Respiration reduces the net photosynthesis.

A plant may be a heliophyte (sun-loving plant) or sciophyte (shade-loving plant) depending on the growth performance of the plant under varying light intensities. The distribution of light in a plant canopy is determined by the transmissibility of the leaf, leaf arrangement and inclination, plant density, plant height, and the angle of the sun. The visible spectrum also influences the orientation of shoots, called phototropism. When the growth of plant is in the direction of light it is known as positive phototropism and when plant growth is away from the direction of sunlight, it is known as negative phototropism or photonasty. With increasing intensity of light, positive phototropism turns into negative phototropism.

Duration of light

Duration of daylight period influences time of flowering of many crop plants. Response of plants to day length is known as photoperiodism. Based on flowering behaviour to photoperiod, plants may be long-day, short-day or day-neutral plants. The long-day plants require long days, usually more than 14 hours of sunshine period and short-day plants require short day length, usually less than 10 hours. The short-day plants are really long-night plants. The flowering of day-neutral plants is independent of day length. The crops or crop varieties which show no response to seasons are called photoinsensitive species or varieties and those which are sensitive to the seasons for flowering behaviour are called photosensitive varieties. Photosensitivity is an adverse character in an intensive cropping system.

Chapter 6

Atmospheric Temperature

Temperature is the physical property of an object that quantitatively expresses the hotness or coldness. Objects of low temperature are cold, while various degrees of higher temperatures are referred to as warm or hot. When a heat transfer path between them is open, heat spontaneously flows from bodies of a higher temperature to bodies of lower temperature. If no net heat flow occurs between two objects, the objects have the same temperature which are then said to be in 'thermal equilibrium'. Temperature plays an important role in all fields of natural and applied sciences. Most of the weather parameters are dependent on it, directly or indirectly. It has important role to play in the development and growth of all living organisms in all stages of their growth.

Atmosphere receives the heat energy from the sun and its temperature increases. Due to different amount of heat energy receipt at different places, the air temperatures at different places also vary. The variation in air temperature basically results into air motion, so as to equalise the energy content of the different regions of the earth. Thus, atmospheric temperature is one of the basic elements of weather and climate.

CONTROLS OF THE AIR TEMPERATURE

The temperature at a given location and time is affected by energy transfer processes including conduction, convection, advection and radiation. The controls of temperature include those factors that bring about spatial variations in temperature. The important factors that determine the temperature of a particular place on the earth's surface are latitude, altitude, distribution of land and water, topography, ocean currents, prevailing wind, cloudiness, precipitation, nature of the surface, and turbulence.

Latitude

Differences in the receipt of solar radiation are the most important cause for temperature variation. Because variation in sun angle and length of daylight depend on latitude, it is the most important factor on air temperature variation. Highest temperatures are generally at the equator and the lowest at the poles.

The temperature decreases with the increase of latitude. Due to earth's spherical shape, the angle of sun's rays on surface of the earth differs from one place to another and temperature changes at every latitude. The band of rays from sun falling vertically over the equatorial latitudes heats up a smaller surface, where the temperature is high. The band of rays falling obliquely over the temperate latitudes heats a larger area (this is due to the curvature of the earth), and there the temperature is low. At polar regions the rays are always slanting and have little heating power.

Altitude

Atmosphere is not directly heated by the sun's rays. First, the sun's rays heat the earth's surface and then, the lower atmosphere is heated by conduction from the earth. Since the direct source of atmospheric heat lies at the surface of the earth, the higher we ascend in the atmosphere, the lower is the temperature. Lower layers of the atmosphere are denser and contain more water vapour, more dust and aerosol particles that absorb a large amount of terrestrial radiation. So the temperature nearer to the earth's surface is higher than that at the upper layers. The permanent snow caps on high mountains, even in low latitudes, indicate the decrease of the temperature with altitude.

Distribution of land and water

Land and water differ in their behaviour when heated. Land surface is heated more quickly than the water surface when subjected to an equal amount of insolation and it also cools more rapidly. It is because of the specific heat of water is much higher than that of land and it requires more heat to increase its temperature. Water being fluid is highly mobile and heat energy is redistributed through turbulence. Heat is absorbed only at the surface on land. But water being more transparent allows more solar radiation to penetrate to a depth of several metres. Evaporation process, which is greater from water bodies, takes up energy as latent heat. All these factors collectively cause water to warm more slowly, store greater quantities of heat and cool more slowly than land. So during the day, the land is hotter than the sea. The heated air over the land becomes lighter and rises. This creates a low pressure area over the land. At this time, the air above the sea is cooler. When this air blows towards the low pressure area over the land, it lowers the temperature of the air on land. This is known as 'sea breeze'. This sea breeze has a moderating influence on coastal regions. During the night, conditions are just the reverse. At this time, the sea remains warmer than the land. So the 'land breeze' blows from the land to the sea, and relieves the cold of the land. The areas, close to the sea have lower daily and annual ranges of temperatures.

Ocean currents

Ocean currents move from one place to another in order to equalise temperatures. Warm ocean currents move from the equatorial regions towards the polar regions as surface currents and cold currents move from polar regions to equatorial regions as deep currents. The warm currents raise the temperature of the nearby coastal regions. Similarly, the cold currents tend to lower down the temperatures of the adjacent land areas.

Prevailing wind

Winds indirectly change the temperature of a place in the neighbourhood of sea. When the prevailing wind is from the oceans, the ocean and ocean currents can exert maximum influence on the adjacent land. The prevailing winds, in temperate latitudes, coming from the land area lower the temperature in winter and raise it in summer. In contrast, the prevailing winds coming from the sea areas raise the winter temperatures, but lower the summer temperatures.

Mountain barriers

Mountain ranges affect the temperature by acting as obstacles to the flow of cold air near the surface and they often set conditions of warm winds.

Clouds

Clouds in the sky also affect the temperature of a place. Thick cloud cover reduces the amount of insolation received at a particular place, thus lowering the day time temperature. During night time, clouds can also absorb outgoing longwave radiation, thus increasing the temperature. The deserts, where sky is mostly cloudfree, experience a high day temperature and a much lower night temperature. In the equatorial areas, because of the abundance of cloud cover, day temperatures are not so high and night temperatures not so low.

Nature of the surface

Different surfaces reflect different amount of incoming solar radiation falling thereon. Dark coloured surface absorbs more insolation than the light coloured ones. Snow covered surfaces reflect a large part of the incident ray, that is why the temperature over snow covered ground remains low even during the daylight hours.

Slope of the land

The effect of slope of the land is not so much noticed in the tropical areas, because in those latitudes the mid-day sun is always high in the sky, and its rays fall more or less vertically over the whole land. But in temperate latitudes, the sun facing slope of mountains that are more or less parallel to the equator gets warmer than the land sloping away from the sun or towards the poles.

This is because the slope facing the sun will get more or less vertical rays, whereas the slope away from the sun will get slanting rays. In northern hemisphere, the southward facing slopes are exposed to the sun for a longer period of time and are in receipt of greater amount of insolation than the northward slopes.

DAILY AND SEASONAL VARIATIONS OF TEMPERATURE

The temperature continuously changes during a day, week, month, year or any period and this change is called periodic temperature variation. Air temperature generally varies from day to night because of changes in the amount of radiation heating earth's atmosphere. Air temperature also changes with the seasons. Summers are warmer than winters except near the equator where temperatures are fairly constant throughout the year. The daily and annual variation in temperature is related to the earth's rotation and revolution around the sun, respectively.

Diurnal temperature variation

The cycle of temperature change during a day is called diurnal temperature variation or daily temperature cycle. A rhythmic rise and fall of air temperature occurs almost every day. The difference between the highest and the lowest temperature of day is the diurnal range. Variation in net radiation (incoming solar radiation minus outgoing terrestrial radiation) is primarily responsible for the particular patterns of rising and falling air temperature over a 24 hour period. At the earth's surface, quantities of insloation and net radiation undergo daily cycles of change because the planet rotates on its polar axis once every 24 hours. The minimum temperature is generally observed early in the morning at about sunrise on any clear day. From sunrise the earth's surface receives solar radiation and incoming radiation is more than outgoing radiation. The maximum temperature in a day is usually recorded at about 14.00–16.00 hours though the maximum amount of solar radiation is received at the noon (*i.e.*, 12.00 hours). This delay in occurrence of maximum temperature is caused by gradual heating of the air by convective heat transfer from the ground which is known as thermal lag or thermal inertia. The day length, condition of the sky, nature of surface, wind speed, humidity and soil composition are the major factors to control the diurnal temperature variation.

Day length: The greater (or shorter) the number of daylight hours, the more (or less) time the earth is subjected to heating. Length of daylight hours is determined by geographic location as well as season.

Condition of the sky: Cloud cover in the sky has a dominant control over the daily ranged temperature. Clouds are good at both absorbing and giving off longwave radiation, and at reflecting shortwave radiation (sunlight). Cloudy and overcast sky moderate the daily range of temperature. When the day is overcast, the passage of incoming solar radiation to the ground is blocked, so that the maximum temperature for the day is reduced. The amount of cloudiness affects the minimum temperature as well.

Nature of the surface: Because of the differential heating of land and water, the places with marine influence, have smaller diurnal range of temperature. On the water surface the maximum temperatures are relatively lower and the minimum temperatures are higher than those on the land surfaces. Therefore, the places situated far away from the oceans have higher diurnal range of temperature.

Wind speed: Winds cause air at differe⁻t levels of the atmosphere to mix. This mixing lessens the difference in temperature between warmer and cooler air, thus decreasing the diurnal temperature range.

Humidity: Lesser the amount of water vapours in the air, greater the amount of heat lost to space by longwave earth radiation. On the other hand, larger amount of water vapour in the air absorbs a large percentage of radiant heat from the earth's surface and minimises the decrease in temperature. Therefore, more humid the air, smaller is the diurnal range of temperature, and drier the air, larger is the diurnal range. That is why higher daily range of temperature is very common in the deserts, and relatively lower range characterises the humid regions.

Soil composition: Heat conductivity of soil increases in direct proportion to its moisture content. In humid soil regions a fraction of insolation received at the ground is expended in evaporating the moisture. This causes the diurnal range of temperature to be higher in dry regions.

Seasonal temperature variation

As the earth revolves around the sun, locations on the surface may undergo seasonal changes in air temperature because of annual variations in the intensity of net radiation. Variations in net radiation are primarily controlled by changes in the intensity and duration of received solar insolation which are driven by variations in day length and angle of incidence. The annual temperature variation gives rise to seasons *i.e.*, summer and winter. The annual temperature range varies greatly from place to place. It reflects the daily increase in insulation from mid-winter to mid-summer and decrease in the same from mid-summer to mid-winter. Usually there is a temperature lag of 30 to 40 days after the period of maximum and minimum insulation.

During June, the northern hemisphere is exposed to more direct sunlight because the hemisphere faces the sun. The same is true of the southern hemisphere in December. However, due to seasonal lag, in the northern hemisphere winter minimum occurs in January and summer maximum in July and *vice versa* in the southern hemisphere. The difference between the highest and the lowest temperature for a given period is known as temperature range. The annual range of temperature may be described as the difference between the temperature of the coldest month and the hottest month. The mean annual range is the difference between the hottest and the coldest mean monthly values. Latitude is the most important factor for annual range. The smaller range occurs near equator and the largest in high latitudes. In the northern hemisphere it is summer from 21st of March to 22nd of September and winter from 23rd September to 20th March and *vice versa* in southern hemisphere.

AIR TEMPERATURE DISTRIBUTION

The atmospheric temperature varies from place to place and time to time, both horizontally and vertically. Distribution of temperature across the latitudes over the surface of the earth is called its horizontal distribution. On maps, the horizontal distribution of temperature is commonly shown by isotherms. Isotherms are lines connecting points that have an equal temperature.

Horizontal air temperature distribution

The rate of change of temperature with a horizontal distance is known as horizontal temperature gradient. Isotherms are generally used to examine the distribution of air temperature over large areas. These are the lines connecting points on a map that have equal temperature. The average global surface air temperature is about 13° C but local averages vary widely. Several factors like latitude and altitude of a place, distribution of land and water on earth's surface, ocean's currents, wind direction and air mass, mountain barriers and topography of a place influence the horizontal distribution of air temperature

Maximum solar energy is received in equatorial region and therefore highest temperatures are observed in equatorial region. As the latitude increases the solar energy received on the earth correspondingly decreases and so also temperature decreases with increase in latitude being the lowest on the pole. Generally temperatures of the places above sea level are converted to their equivalents at sea level before connecting points of equal temperature. In other words, the temperature data are plotted on maps only after necessary corrections to eliminate the effect of altitude are done. All points through which an isotherm passes have identical temperature for the time period mentioned. The isothermal lines generally trend east-west giving an evidence of the dominant control of latitudes on surface distribution of temperature. Temperature does not fluctuate as much over water as over land. The coldest temperature in winter and the highest temperature in summer are found over the continents. The prevailing wind direction has a direct influence upon the average temperature of an area. The winds coming from ocean to land or *vice versa* have considerable influence on temperature of a locality. Mountain ranges influence horizontal temperature distribution by restricting movement of air masses.

The highest and the lowest recorded natural temperature on earth is 57.8° C (at El' Azizia in Libya on September 13, 1922) and –89.4° C (at Vostok in Antartica on July 21, 1983), respectively. World's highest (34.4° C) and lowest (–57.2° C) mean annual temperature have been observed at Dallol (Ethiopia) and Sovietskaya (Antartica), respectively. The record cold temperature for an inhabited area was –68° C at Oymyakon (Siberia) on February 6, 1933. Maximum difference between the lowest and the highest recorded temperatures from the same location is 101.8° C at Verkhoyansk (Russia).

Vertical profile of air temperature

Several factors like physical properties and albedo of the surface, intensity and duration of solar energy, depletion of energy received on the surface by scattering, reflection and absorption, and transport of heat by advection, convection and turbulence influence the vertical distribution of air temperature.

The rate of change of temperature over a vertical distance is known as vertical temperature gradient or **lapse rate**. It varies with height, location and season. A positive lapse rate indicates a decrease of temperature with height. A negative lapse rate indicates an increase of temperature with height which is called **temperature inversion**. The average lapse rate or decrease of temperature with altitude is known as normal lapse rate. In troposphere, the normal lapse rate is about 6.5° C/km. However, in between 4 to 6 km and 6 to 8 km the average rate of temperature decrease is about 6 and 7° C/km, respectively. These normal lapse rates represent average of many observations at different places and times, and maybe different from the actual lapse rate (or environmental lapse rate), indicated by single observation over a given location at a given time. When there is no change in temperature with altitude, the lapse rate is termed isothermal. An isothermal layer is defined as a vertical column of air having a constant temperature with height. Isothermal layers often occur in the low levels of the troposphere.

Adiabatic process

The process of change in temperature without any exchange of heat of a system with its environment is called adiabatic process. It is the opposite of a diabatic process, in which heat exchange occurs. Adiabatic changes in air temperature occur due to changes in pressure of the air while not adding or subtracting any heat. Due to the increase in pressure on a descending parcel of air, the parcel's volume decreases and its temperature increases, thus increasing the internal energy. Adiabatic heating occurs when the pressure of an air parcel is increased from work done on it by its surroundings. Similarly, adiabatic cooling occurs when the pressure of an air parcel is decreased due to the work done on it by its surroundings. When the pressure applied on a parcel of air decreases, the air in the parcel is allowed to expand; as the volume increases, the temperature falls and internal energy decreases. The rate at which the temperature changes as air rises or falls is called adiabatic lapse rate.

If a parcel of air is lifted upward, its pressure is decreased, and its temperature falls due to the expansion. If the air is dry or unsaturated (*i.e.*, relative humidity below 100%) and the process is adiabatic, the rate of temperature fall is 10° C/km of lift. If that parcel descends again to higher pressure, its temperature increases at the rate of 10° C/km. This is known as **dry adiabatic lapse rate**.

When a mass of air is lifted, it cools at the dry adiabatic lapse rate of 10° C/km as long as it remains unsaturated. If the original moisture is being carried along with the mass as it ascends and it cools to its saturation temperature, the relative humidity reaches 100%. Condensation takes place with further cooling. For each gram of water condensed, about 597 calories of heat is liberated. This latent heat of condensation is absorbed by the air, and the adiabatic cooling rate is decreased to 5° C/km instead of 10° C/km. The process during the saturated expansion of the air is called saturation adiabatic, the moist adiabatic or **wet adiabatic lapse rate**.

The reason for the difference between the dry and moist adiabatic lapse rate values is that latent heat is released when water condenses, thus decreasing the rate of temperature drop as altitude increases. This heat release process is an important source of energy in the development of thunderstorms.

Stability of air

Adiabatic temperature change is an important factor in determining the stability of the air. Stability or atmospheric stability is a term used to describe the convective properties of an air mass. The tendency for air to rise or fall through the atmosphere may cause variation in air stability. Stable air has a tendency to resist movement. On the other hand, unstable air will easily rise. The tendency for air to rise or fall depends on the adiabatic and environmental lapse rates. Stable atmospheric conditions prevail when the environmental lapse rate is less than the saturated adiabatic rate. Air is unstable when the environmental lapse rate is greater than the dry adiabatic rate. Under these conditions, a rising parcel of air is warmer and less dense than the air surrounding it at any given elevation.

When a cool air mass moves over a warmer region, the warm surface will heat the parcel of air from below. The start of this convective process will create an air mass that is less stable as the temperature can change quite rapidly. If a warm air mass moves over a cold surface, the air becomes more stable as the convective processes in the air slow. The more stable the air mass, the less likely a region will receive precipitation.

TEMPERATURE INVERSION

The temperature inversion is a condition in which the temperature of the atmosphere increases with altitude in contrast to the normal decrease with altitude. Under certain conditions, the normal vertical temperature gradient or lapse rate is inverted so that the air is cooler near the surface of the earth. The inversions are very common in winter over continents. Inversion phenomenon influences the condensation, precipitation and atmospheric visibility. Inverted layers act as cover for rising air currents, thus an inversion can lead to pollution such as smog being trapped close to the ground, with possible adverse effects on animal health and crop growth. During a temperature inversion, air pollution released into the atmosphere's lowest layer is trapped there and can be removed only by strong horizontal winds. Because high-pressure systems often combine temperature inversion conditions and low wind speeds, their long residency over an industrial area usually results in severe smog. A temperature inversion can also suppress convection by acting as a 'cap'. If this cap is broken for any reason, convection of any moisture present can then erupt into violent thunderstorms. It also affects the weather of a particular locality by controlling vertical distribution of temperature in the atmosphere. Temperature inversion can sometimes result in freezing rain in cold climates. The ideal conditions of temperature inversion are absence of winds, clear sky, long winter nights, cold dry air and snow covered earth.

In mountain regions, at times, the temperature is less in valleys than at greater heights. It occurs particularly on calm, cold winter nights, when the sky is clear and the air very dry. Because of these conditions, the heat from the earth's surface and the lower layer of the atmosphere escapes rapidly back into space, making the upper slopes warmer. The cold is further increased by the sinking of the cool air from the mountain sides. Thus, there is a reversal in the vertical distribution of temperature. This phenomenon is very common in the mid-latitude regions. There are four types of inversion depending on conditions under which they occur.

Ground or surface temperature inversion

A ground inversion develops when air is cooled by contact with a colder surface until it becomes cooler than the overlying atmosphere; this occurs most often on clear nights, when the ground cools off rapidly by radiation. If the temperature of surface air drops below its dew point, fog may result. Topography greatly affects the magnitude of ground inversions. If the land is hilly, the cold air formed on the higher land surfaces tends to drain into the hollows, producing a larger and thicker inversion above low ground and little or none above higher elevations. This type of inversion in lower and middle latitudes occurs during cold nights and gets destroyed during daytime.

Intermountain valley inversion

Sometimes, the temperature in the lower layers of air increases instead of decreasing with elevation. This happens commonly along a sloping surface. Here, the surface radiates heat back to space rapidly and cools down at a faster rate than the upper layers. As a result the lower cold layers get condensed and become heavy. The sloping surface underneath makes them move towards the bottom where the cold layer settles down as a zone of low temperature while the upper layers are relatively warmer. This kind of temperature inversion is very strong in the middle and higher latitudes. It can be strong in regions with high mountains or deep valleys also.

Subsidence or upper surface temperature inversion

A subsidence inversion develops when a widespread layer of air descends. The layer is compressed and heated by the resulting increase in atmospheric pressure, and as a result the lapse rate of temperature is reduced. If the air mass sinks low enough, the air at higher altitudes becomes warmer than at lower altitudes, producing a temperature inversion. Subsidence inversions are common over the northern continents in winter and over the subtropical oceans; these regions generally have subsiding air because they are located under large high-pressure centres. This temperature inversion is called upper surface temperature inversion because it takes place in the upper parts of the atmosphere.

Frontal or advectional inversion

A frontal inversion occurs when a cold air mass undercuts a warm air mass and lifts it aloft; the front between the two air masses then has warm air above and cold air below. This kind of inversion has considerable slope, whereas other inversions are nearly horizontal. In addition, humidity may be high, and clouds may be present immediately above it. This type of inversion is unstable and is destroyed as the weather changes.

EFFECT OF AIR TEMPERATURE ON PLANT GROWTH

Air temperature is the most important climatic parameter that affects the growth and productivity of plants, depending on whether the plant is a warm season or cool season crop. Temperature influences distribution of crop plants and vegetation. It affects leaf production, flowering and fruit setting. Physical and chemical processes within the plants are governed by air temperature. It also affects the stability of enzymatic systems in the plants.

Cardinal temperature

Below a certain minimum and beyond maximum value of temperatures, plant growth ceases. Between these limits, there is an optimum temperature at which growth proceeds with the greatest rapidity. These three points of temperature are the cardinal temperatures for a given plant. The cardinal temperatures are known for most plant species, at least approximately. Cool season crops have low cardinal temperature. For typical cool season crops, the minimum temperature ranges between 0 and 5° C, optimum temperature ranges between 25 and 31° C, and maximum temperature ranges between 31 and 37° C. Similarly, for hot season crops, the minimum temperature ranges between 15 and 18° C, optimum temperature ranges between 31 and 37° C, and maximum temperature ranges between 44 and 50° C. Cardinal temperatures may also vary with the stage of development.

Table 6.1 Cardinal temperatures for germination of some crops

Plant	Cardinal temperature (in degree Celsius)		
	Minimum	Optimum	Maximum
Rice	10–12	30–32	36–38
Wheat	3–4.5	25	30–32
Maize	8–10	32–35	40–44
Barley	3–4.5	20	38–40
Sorghum	8–10	32–35	40
Oats	4–5	25	30
Peas	1–2	30	35
Lentil	4–5	30	36
Pumpkin	12	32–34	40
Sugarbeet	4–5	25	28–30
Carrot	4–5	8	25
Tobacco	13–14	28	35

(*Source*: Bierhuizen, 1973)

Thermoperiodism

The response of plants to diurnal or seasonal temperature fluctuation is known as thermoperiodism. All plants are subject to particular temperature extremes and fluctuations which they can tolerate. Thermoperiods which exceed these extremes can degrade plant tissue and kill the organism. Different growth phases of a crop plant may be influenced by the thermoperiods to which it is exposed. Particularly, it has been found that diurnal thermoperiods can play a significant role in the germination of seeds. The physical shape or morphology of a plant can also be affected by the differences in day and night temperature. The diurnal temperature differences influence internode length, plant height, leaf orientation, shoot orientation, chlorophyll content, lateral branching and petiole and flower stalk elongation.

Daytime temperatures provide the greatest heat load, but night temperatures are also critical for many plant growth mechanisms, especially reproductive structures. Night temperatures are critical for controlling respiration rate. Rates of both photosynthesis and respiration rise with increase in air temperatures. As temperatures reach the upper growing limits for the crop, the rate of food use by respiration may exceed the rate at which food is manufactured by photosynthesis. Heat stress problems also make the plant more susceptible to pests and other environmental problems. A number of pathogenic fungi are more effective in attacking trees when the host is under heat stress.

Low temperature injury

Low temperature affects several aspects of crop growth, *viz.*, survival, cell division, photosynthesis, water transport, growth and finally yield.

Suffocation: During winter, formation of a thick cover of ice or snow on the soil surface prevents the entry of oxygen and small crop plants may suffer. Certain toxic substances accumulate in roots and crowns because of low diffusion of carbon dioxide.

Heaving: The injury to a plant is caused by lifting upward of the plant from its normal position in temperate regions where snowfall is a common phenomenon. Sometimes the roots are pushed completely above the soil surface. It is difficult for the roots to become firmly established again and the plants may die because of this mechanical damage and desiccation.

Chilling injury: Many fruits, vegetables, and ornamentals of tropical or subtropical origin are sensitive to low temperatures. These crops are injured after a period of exposure to chilling temperatures below 10 to 15° C but above their freezing points (0° C). Certain horticultural crops of temperate origin are also susceptible to chilling injury. These temperate crops, in general, have lower threshold temperatures, < 5° C. At these chilling temperatures, the tissues weaken because they are unable to carry on normal metabolic processes. Various physiological and biochemical alterations and cellular dysfunctions occur in chilling-sensitive species in response to chilling stress.

When chilling stress is prolonged, these alterations and dysfunctions will lead to the development of a variety of chilling injury symptoms such as surface lesions, internal discolouration, water-soaking of the tissue, and failure to ripen normally.

Freezing injury: This is generally caused in plants growing in temperate regions. When the plants are exposed to very low temperature, water freezes into ice crystals in the intercellular spaces. With further fall in temperature, water is withdrawn from the cells, resulting in increase in the size of ice crystals in the intercellular spaces. Ice within the cells cause injury by mechanical damage and brings about dehydration of cell contents. Thus, plant parts or entire plant may be killed or damaged. Tender plants, like tomato, die when they are exposed to frost.

Heat stress syndrome

Heat stress syndrome often is defined as symptoms of irreversible damage to plant function or development where temperatures are hot enough for sufficient time. In addition, high temperatures can increase the rate of reproductive development, which shortens the time for photosynthesis to contribute to fruit or seed production.

In other words, heat stress syndrome is a series of metabolic dysfunctions and physical constraints that pileup inside plants and become impossible for a plant to adjust, avoid or correct. The duration of hot temperatures for plants must not exceed the plant's ability to adjust, avoid, or repair problems. The ten step heat stress syndrome sequence is given below.

1. decrease in photosynthesis
2. increased respiration
3. closing down of photosynthesis; closed stomata stop CO_2 capture and increased photo-respiration
4. major slowdown in transpiration (cooling process loss and internal temperature increase)
5. cell membrane leakage
6. continued physical water loss
7. growth inhibition
8. plant starvation through rapid use of food reserves, inefficient food use, and inability to call on reserves when and where needed
9. toxins generated through cell membrane releases and respiration problems
10. membrane integrity loss and protein breakdown

High temperature adversely affects mineral nutrition, shoot growth and pollen development resulting in low yield. Heat injury in plants includes scorching of leaves and twigs, sunburn on branches and stems, leaf senescence and abscission, acute leaf death, and shoot and root growth inhibition. The critical temperature above which plants get killed is called thermal death point. Many annual crop plants may be killed at the temperature above 50° C.

GROWING DEGREE DAYS

Temperature plays a very important role in the developments of plants. As a crop plant develops over a growing season, its development is closely related to the daily accumulation of heat. A certain amount of heat is required to provide enough energy for the plant to move to the next development stage. The amount of heat required to move to the next development stage remains constant from year to year, but depending on weather conditions, the amount of actual time may vary.

When an organism's growth rate is predominantly under temperature control its rate of development can be characterised using a system called growing degree days (GDD) or heat units (HU) or thermal units (TU). A degree day or heat unit is the departure from the mean daily temperature above the minimum base temperature. This GDD concept assumes that

1. Each organism has a minimum base temperature or threshold temperature or reference temperature below which no growth occurs.
2. There is a direct linear relationship between plant growth and temperature above the base temperature.
3. Growth and development are closely related to daily temperature mean accumulations above the base temperature.

These base temperatures have been determined experimentally and are different for each species. Tropical plants have higher base temperature while temperate plants have lower base temperature. It varies from 4.5 to 12.5° C for most of the crop plants.

Table 6.2 Threshold or base temperatures of crops

Crops	Threshold temperature (in °C)
Rice	10–12
Wheat, barley,	4.4
Potato, oat, sugar beet	5.0
Sorghum, maize, groundnut, soybean, tomato	8–10
Tobacco	13–14

To calculate GDD, the mean temperature for the day is first determined by adding the maximum and minimum temperatures for the day and dividing it by 2. The base temperature is then subtracted from the mean temperature to give a daily GDD. If the daily GDD calculates to a negative number it is made equal to zero. Each daily GDD is then added up (accumulated) over the growing season.

$$GDD = (T_{max} + T_{min})/2 - T_{base}$$

where, T_{max} and T_{min} are maximum and minimum daily temperature of the day and T_{base} is the base temperature for the crop plant.

A modified GDD may be calculated with some temperature adjustments. If the observed daily maximum and minimum temperatures are beyond and below the temperatures where no growth occurs in a crop species, the daily maximum and minimum temperatures for calculation of the GDD for that crop are reset to the temperature limits where no growth occurs, instead of the actual observed temperatures. For example, in case of maize if the daily maximum temperature is greater than 30° C, then it is reset to 30° C and if the daily minimum temperature is less than 10° C, then it is reset to 10° C. These modified temperatures are averaged and compared to the base or reference temperature, assumption being that development in maize does not take place beyond 30° C.

Growing degree days have many practical applications in crop management. This concept can be used to assess the suitability of a region for production of a particular crop. Accumulated GDD, calculated by summing GDDs for each day during a period, are used to estimate the growth stages of crops and harvest maturity of crops, weeds or even life stages of insects; predict maturity and cutting dates of forage crops; predict best timing of fertiliser or pesticide application; estimate the heat stress on crops; plan spacing of planting dates to produce separate harvest dates; schedule vegetable harvesting according to their level of tenderness. The GDD can also be used to predict the duration, under natural conditions, between two developmental stages for a parasite used in a crop for biological control of a pest; anticipate insect development with a view to better phytosanitary control; and assess sporulation intensity for cereal fungi.

Limitations of GDD

Though the 'growing degree days' concept is simple and very useful, it also has many limitations.

1. It assumes linear relationship between temperature and growth and more weightage is given to higher temperature, although in many crops temperature above 30° C may have detrimental effects on growth.
2. The daily range of temperature is not considered. In fact, this may carry more significance than the mean daily temperature.

3. A single threshold temperature is used throughout the life cycle of a crop. But there may be different threshold temperatures for different growth phases of a crop. This differentiation is not considered in calculating the accumulation of GDD.
4. The effects of topography, altitude and latitude on crop growth are not considered.
5. The effect of soil fertility and other edaphic factors on crop growth cannot be explained by this concept.
6. The effect of other climatic parameters like solar radiation, wind, precipitation is not considered.
7. The effect of biotic factors like insects, diseases and weeds on crop growth is not considered.
8. Crop leaf or canopy temperature may have more impact on crop growth rather than the temperature recorded in the observatory.

Modifications in GDD

Many researchers have suggested improvements in the GDD concept by inclusion of other weather parameters along with temperature. However, these modified indices are not free from limitations.

Photothermal unit: This is calculated by multiplying GDD with average hours of day length. This is based on the assumption that plant growth is controlled by length of day and night and heat requirements vary with photoperiod.

Heliothermal unit: This index is calculated by multiplying GDD with actual bright sunshine hours recorded in the observatory by a sunshine recorder.

Energy degree unit: This is calculated by multiplying GDD with daily solar radiation directly by radiometer or calculated using empirical equations.

Soil moisture degree day: This is calculated by multiplying GDD with the per cent ratio of available soil moisture and maximum soil moisture content.

Heat use efficiency: This is calculated by dividing total dry matter production by GDD.

Heliothermal use efficiency: This can be calculated by dividing total dry matter production by heliothermal unit.

Radiation use efficiency: This index can be calculated for the entire growing season or for each phenophase of the crop by dividing total dry matter production by intercepted photosynthetically active radiation.

SOIL TEMPERATURE

Soil temperature is an important environmental parameter for plant growth and development. It has direct effects on microbial growth and development, organic matter decay, seed germination, root development, and water and nutrient absorption by roots. In general, the higher the temperature the faster these processes occur. Variation in soil temperature is much more pronounced because of varying characteristics and composition of soil.

The soil gains heat by absorption of solar radiation and longwave counter radiation from the atmosphere, downward transfer of sensible heat and horizontal transfer of heat from the surroundings. On the other hand, soil loses heat by longwave radiation to the atmosphere, upward transfer of sensible heat to air when the soil surface is hotter than air and with evaporation and horizontal transfer of heat from the column to surroundings.

Factors affecting soil temperature

The temperature of the soil is determined by exchange of heat, then temperature will increase or decrease depending on heat capacity of the soil. Both environmental and soil factors contribute to changes of soil temperature.

Solar radiation: The amount of solar radiation available at any given location and point of time is directly proportional to soil temperature. Even though a part of total net radiation available is utilised in evapotranspiration and heating the air, a relatively substantial amount of solar radiation is utilised in heating up of soil (ground heat flux) depending upon nature of surface. Radiation from the sky contributes a large amount of heat to the soil in areas where the sun's rays have to penetrate the earth's atmosphere very obliquely.

Evaporation: The greater the rate of evaporation the more the soil is cooled. This is the reason for coolness of moist soil in windy conditions.

Condensation: Whenever water vapour from soil depths or atmosphere condenses in the soil, its heat increases noticeably. Freezing of water generates heat.

Rain fall (precipitation): Depending on its temperature, precipitation can either cool or warm the soil.

Vegetation: A bare soil quickly absorbs heat and becomes very hot during the summer and become very cold during the winter. Vegetation acts as an insulating agent. It does not allow the soil to become either too hot during the summer or too cold during the winter.

Wind: Air convection or wind is necessary to heat up the soil by conduction from the atmosphere. The mountain and valley winds influence the soil temperature.

Aspect and slope: These factors are important in determining soil temperature in temperate regions. In the northern hemisphere the south facing slope is always warmer than a north facing slope or plain as it receives comparatively more insolation per unit area.

Exposure and latitude: The exposure is of little importance in tropics near the equator due to high elevation of the sun. It is of more significance in the middle latitudes where the elevation of the sun is lower. The effect of latitude is primarily on the distribution of solar radiation on the surface of the earth due to the angle which the sunrays make with the earth.

Tillage and tilth: Shallow tillage and creating 'dust mulch' reduces the heat flow between the surface soil and subsoil. Since the soil mulch has a greater exposed area and the capillary connection with moist layers below is broken, the cultivated soil dries up quickly by evaporation and has greater temperature amplitude as compared with uncultivated soil. In the noon, the air 2 - 3 cm above the tilled soil is hotter by 5 to 10° C than that of an untilled soil. However, at night loosened ground is colder and more liable to frost than the uncultivated soil.

Moisture content: Moisture has an effect on heat capacity and heat conductivity. A soil with higher moisture content is cooler than dry soil. Availability of optimum soil moisture maintains the soil temperature at required level.

Soil texture: Soil texture affects the thermal conductivity of soil. It decreases with reduction in particle size. Sandy soils warm up more rapidly than clay soils, because of lower heat capacity of sandy soils. A soil with rough surface absorbs more solar radiation than one with a smooth surface. Radiational cooling at night is greater in light soils than in heavy soils.

Organic matter content: Organic matter reduces the heat capacity and thermal conductivity of soil. Dry organic soils are warmer than mineral soils in summer and are cooler in winter.

Colour of the soil: A black coloured soil absorbs more heat than light coloured soils. Hence black colour soils are warmer than light coloured soils during bright days. At the same time, black coloured soils also tend to lose energy at a higher rate. The dark coloured soils emit more heat during the night and thus, allow a greater daily variation in temperature. Albedo, *i.e.*, the reflection factor for shortwave radiation is related to soil colour. Albedo is the ratio of amount of visible light reflected by a body to the amount incident upon it. It is also called reflected radiation and expressed in percentage. Some of the albedo values of natural surfaces for total shortwave radiation are given below.

Table 6.3 Albedo values of natural surfaces for total shortwave radiation

Surface	Albedo (in %)
Fresh snow	75–95
Water	3–10
Sand	35
Granite	15
Dark clay, wet	2–8
Dark clay, dry	16
Sand, wet	18
Sand, dry	18
Bare fields	12–25
Wet ploughed fields	5–24
Stubble fields	15–17
Grass, green	16–27
Grass, dried	19–19
Grain crops	10–15
Desert- mid day	15
Forests	5–20

(*Source*: Ghildyal and Tripathy, 1987)

Thermal properties of soil

The heat is continuously moving in to or out of the soil and thermal energy is being continuously redistributed in the soil. Heat will not flow under isothermal condition. Thermal properties relevant to the transfer of heat through a medium and its effect on the average temperature in the medium are the specific heat, heat capacity, thermal conductivity and thermal diffusivity.

Specific heat: It is the amount of heat required to raise the temperature of one gram of substance by 1° C. The values for most minerals present in the soil are between 0.18 and 0.20 cal and that of water is 1 cal.

Heat capacity: It is the amount of heat required to raise the temperature of one cubic centimetre substance by 1° C. Specific heat multiplied by soil mass is the heat capacity of soil and it is the summation of specific heat of soil constituents. Most soils have a heat capacity in the range of 0.3 to 0.6 cal/cm^3 and that of water is 1 cal/cm^3.

Thermal conductivity: It is the ability of the substance to transfer heat from molecule to molecule. Thermal conductivity is defined as the quantity of heat flowing in a unit time through a 1 cm^2 cross section of the soil in response to a temperature gradient of 1° C per cm of depth. It varies with porosity, moisture content and organic matter content of soil.

Thermal diffusivity: The thermal diffusivity of a material is a measure of how fast the material temperature adapts to the surrounding temperature. It is the change in temperature in °C that occurs in one second when the temperature gradient changes by 1° C per cubic cm. In other words, it is the ratio of the thermal conductivity to heat capacity.

Table 6.4 Thermal properties of soil components

Material	Density (g/cm^3)	Specific heat (Cal/g/°C)	Thermal conductivity (Cal/cm/sec/°C)	Thermal diffusivity (cm^2/sec)
Clay	1.8	0.8	0.0029	0.002
Humus soil	0.3	0.3	0.00027	0.003
Wet sandy soil	1.6	0.4	0.0064	0.01
Air	0.0012	0.24	0.00005	0.1667
Water	1.0	1.0	0.0014	0.0014
Ice	0.917	0.505	0.005	0.0108

(Source: Rosenberg, 1974)

Soil temperature fluctuations

Soil constitutes a major storage for heat acting as a sink of energy during the day and source to the surface at night. In annual terms, the soil stores energy during the warm season and releases it to air during the cold portions of the year. The diurnal and annual soil temperature fluctuations are caused mainly by variations in air temperature and solar radiation. Soil temperature changes more slowly than the air temperature, so there is always a lag time between the extremes of air temperatures and soil temperatures. Because of daily temperature fluctuations, the soil could be cooler than the air in the day time and warmer than the air in the night time.

Diurnal soil temperature variations are more pronounced only in the surface soil. The deeper the soil, the more constant the temperature will be. At a depth of 30 cm they hardly exceed 3° C, at 60 cm they may reach about 1° C, while at 1 m depth diurnal temperature fluctuations are almost nil (Deacon, 1969).

The annual variations depend upon the onset and recession of rains and the local hydrological conditions affecting the water table depth, which modify the moisture regime of the soil profile. Annual soil temperature variations occur much deeper into the soil. When the plant canopy is fully developed the seasonal variations are smaller. In winter, the depth to which the soil freezes depends on the duration and severances of the winter. In summer, the soil temperature variations are much more than that in winter in tropics and subtropics.

Soil temperature and plant growth

Soil temperature is an important environmental parameter for plant growth and development by influencing several plant processes.

1. All crops practically slow down their growth below the soil temperature of about 9° C and above the soil temperature of about 50° C.
2. Water absorption in crop plants is slow both at low and high temperatures. Slower elongation of roots, increase in viscosity of water, increased protoplasm viscosity and decrease in permeability of roots are the major causes for low water absorption under low soil temperature. At high temperature, decrease in water absorption is due to faster maturation and root injury. Soil moisture intake by the plants stops when the soil temperature is below 1° C.
3. Soil temperature controls the nutrient availability to the crops. Nutrient absorption increases with increase in soil temperature up to certain level, especially in the case of anionic forms of nutrients. Extreme low temperature impede intake of nutrients. The optimum soil temperature for nitrification is about 32° C. Optimum soil temperature for nodule growth is 24° C. Nodule size decreases during summer season when the crop is subjected to soil moisture stress.
4. Extreme low and high temperature influences the soil microbial population and rate of organic matter decomposition. Most of the soil organisms function best at an optimum soil temperature of 25 to 35° C.
5. Excessive high temperature is harmful to roots and also causes lesions of stem. Root system becomes succulent, slender and white in many crops in tropics and subtropics in low temperature and branch sparsely.
6. Soil temperature plays an important role in seed germination. Adequate soil temperatures for germination range widely for different crops. Low soil temperature may lead to delayed and poor seed germination.
7. Soil temperature is more important for the tuber crops. High soil temperature causes degeneration of tubers in potato in tropics. The optimum soil temperature for this crop is 17° C. There may not be any tuber formation if the soil temperature is above 29° C.

Chapter 7

Atmospheric Humidity

Water is the only substance that exists in the atmosphere in all the three phases, *i.e.*, solid, liquid and gas. The moisture on the earth's surface is in a constant phase change. The moisture content of the atmosphere is about 0.035% of all fresh water. The atmosphere receives its supply of moisture from the earth's surface through evaporation from oceans, lakes, rivers, damp soil, *etc.* and through transpiration from vegetation. Water vapour is the gaseous state of water and is invisible.

The amount of water vapour present in the atmosphere is called atmospheric moisture or humidity. It is a measure of the amount of water vapour dissolved in the air, not including any liquid water or ice falling through the air. While humidity itself is a climate variable, it also interacts strongly with other climate variables. The humidity is affected by winds and by rainfall. Moisture content is determined by the local rate of evaporation, air temperature and the horizontal transport of moisture. Humidity indicates the likelihood of precipitation, dew or fog. It also plays a major role in the heat budget and the weather phenomenon of the atmosphere.

DISTRIBUTION OF WATER VAPOUR IN THE ATMOSPHERE

In the climate system water exists in its various phases (solid, liquid or vapour) and is in constant movement among the subsystems. The movement of water between and within the atmosphere and the earth is referred to as hydrologic cycle. Water vapour is the link between the surface and the atmosphere in the hydrologic cycle. The global hydrosphere consists of a series of reservoirs interconnected by water cycling in various phases. These reservoirs are oceans, ice sheet and glaciers, terrestrial water (river, soil moisture, lake and ground water), the biosphere (water in plants and animals) and the atmosphere.

Water cycling involves evaporation, the transport of water vapour in the atmosphere, condensation, precipitation and terrestrial runoff. Evaporation can occur directly from oceans or from water bodies on land surfaces (such as lakes and rivers). It can also occur indirectly through plants *via* the process called transpiration. Once in the atmosphere, water vapour can be transported horizontally and vertically by the three-dimensional circulation of the atmosphere and may condense to form liquid water or ice crystals in clouds. The cycle is completed when water returns to the earth's surface in various forms of precipitation such as rain or snow fall. This cycle is closely tied to atmospheric circulation and temperature patterns.

Water vapour is constantly being cycled between the atmosphere and the earth's surface. The input rate (the rate at which water vapour enters the atmosphere from the ground through evaporation) and output rate (the rate at which water moves from the atmosphere to ground through precipitation) are equal. The average length of time that individual molecules of water remain in the atmosphere is called the residence time. It can be found by dividing the mass of the substance in the atmosphere (in kilograms) by the rate (in kg per year) at which the substance enters to and exists in the atmosphere. The total volume of atmospheric water vapour at any instance over the globe is about 14000 km^3, which is 0.001% of the hydrosphere.

Although water vapour is lighter than air, the combination of higher temperatures and the ocean source tends to restrict it to the lower atmosphere. The decrease of water vapour with elevation is quite important in many meteorological processes. In fact, most of the water vapour is contained within the lower 5 km of the atmosphere (> 90%) and more than half is actually contained within the lowest 1.5 to 2 km. Less than 5–6% of the water is above 5 km, and less than 1% is in the stratosphere.

Table 7.1 Decrease of water vapour with elevation

Height (km)	Water vapour content in the atmosphere (%)
0	1.30
1	1.00
2	0.69
3	0.49
4	0.37
5	0.27
6	0.15
7	0.09
28	0.05

(*Source*: Donn, 1975)

MEASUREMENT OF ATMOSPHERIC MOISTURE

There are several different methods for defining the amount of water vapour in the atmosphere. They are absolute humidity, specific humidity, mixing ratio, vapour pressure, relative humidity and dew point temperature. The definition used depends on the application. Each humidity measure is controlled to some degree by air temperature.

Absolute humidity

Absolute humidity is the weight of water vapour per unit volume of air, usually measured in units of grams of water vapour per cubic metre of air. Absolute humidity ranges from 0 grams per cubic meter in dry air to 30 grams per cubic meter when the vapour is saturated at 30° C. Absolute humidity is not often used to express the moisture content of air because it is sensitive to changes in both the atmospheric temperature and pressure.

Specific humidity

Specific humidity is measured as the weight of water vapour in the air per unit weight of air including the water vapour. Technically speaking, this is dimensionless as it is the mass of water vapour to the mass of dry air. So it is expressed as kg/kg. However, the mass of water vapour is much less than the value of the mass of dry air and most commonly meteorologists use g/kg. Given that weight is not significantly influenced by temperature or atmospheric pressure, specific humidity is much more useful as a measure of humidity. In general it is greater in the tropics (16 g/kg) than in the polar regions (4 g/kg). This definition is used when one is interested in the actual amount of water that is in the atmosphere as a gas. The specific humidity remains constant as long as no water vapour molecules is added to or taken out from the volume of air.

Mixing ratio

Another measure very similar to specific humidity is the mixing ratio. The mixing ratio is the weight of water vapour per unit weight of dry air. Because the atmosphere is made up of so little moisture by volume, the mixing ratio is virtually the same as the specific humidity. Changing the temperature of the air parcel does not affect the parcel's mixing ratio.

Vapour pressure

Humidity is not only measured as a weight, but also by the pressure it creates. Water vapour exerts pressure due to its weight on unit surface area like any other gases and is known as vapour pressure. It is a partial atmospheric pressure. The vapour pressure exerted due to actual amount of water vapour present in atmosphere is known as actual vapour pressure. Vapour pressure, like atmospheric pressure, is measured in millibars and is relatively insensitive to volumetric expansion or temperature. The saturation vapour pressure is simply the pressure that water vapour creates when the air is fully saturated. The amount of water vapour required for saturation of air depends on atmospheric temperature. With temperature increase, the saturation vapour pressure increases exponentially. If the atmosphere is said to be unsaturated, it means that it can hold moisture further till it is saturated. The difference between the saturation vapour pressure (the moisture holding capacity of air) and its actual water vapour pressure is termed saturation vapour pressure deficit.

Relative humidity

Relative humidity is the ratio of the amount of water vapour actually in the air compared to the amount of water vapour required for saturation at that particular temperature and pressure. It is expressed as a percentage. Increasing or decreasing the amount of water vapour in the air changes the relative humidity. A change in temperature will also bring about a change in relative humidity. The relative humidity of a parcel of air will change if either its temperature or its mixing ratio is changed. In general relative humidity varies inversely with temperature during the day, tending to be higher at night. Relative humidity is often mentioned in weather forecasts and reports, as it is an indicator of the likelihood of precipitation, dew, or fog.

Because relative humidity is based on the air's water-vapour content, as well as the amount of moisture required for saturation, it can be changed in either of two ways. It can be changed by the addition or removal of water vapour through evaporation and condensation processes. A change in air temperature also results in change in relative humidity, but in a negative way. In nature, change in air temperature is related to net solar radiation, horizontal movement of air over the surface and vertical rising and expanding of air.

The maximum relative humidity is recorded at the equator, as the air can hold more moisture there due to high surface air temperature. A gradual decrease is seen towards the subtropical high-pressure belt. Again a gradual increase in relative humidity is seen due to lower temperatures from the latitudes 40 degree onwards to the poles. The relative humidity is lower over the northern hemisphere when compared to that of the southern hemisphere due to difference in distribution of land and ocean. The most humid cities on earth are generally located closer to the equator, near coastal regions.

The diurnal variation in relative humidity is approximately inverse to that of temperature. The maximum relative humidity is observed at about sunrise and the minimum relative humidity is observed within 14 to 15 hours. The annual variation of relative humidity is largely depended upon the locality. At regions where the rainy season is in summer and winter is dry, the maximum relative humidity occurs in summer and the minimum in winter and at other regions the maximum relative humidity occurs in winter. Over the oceans relative humidity reaches the maximum in summer.

Dew point

Since relative humidity is easily changed by temperature or amount of water vapour, meteorologists use another index of humidity, that is, dew point which is not so easily changed. Dew point is the temperature to which the air would have to be cooled, without changing pressure or moisture content, to reach saturation. High dew points indicate a large amount of moisture in the air. Low dew points indicate small amounts of moisture in the air. When the dew point is below zero degrees Celsius it is called the frost point. Dew points are directly related to the amount of moisture in the air, so it is a better indicator of humidity.

Dew point is normally reached when a mass of air has a relative humidity of 100%. At dew point temperature, water vapour changes into liquid or solid usually forming rain, snow, frost or dew at constant pressure, without addition or removal of vapour. This happens in the atmosphere as a result of cooling through a number of different processes. Thus, the temperature at which the saturation vapour pressure is equal to actual vapour pressure can also be known as dew point temperature. If the dew point temperature is close to the air temperature, the air becomes saturated. The dew point temperature can be calculated using hygrometric tables, based on dry and wet bulb thermometer readings at a given location. If the dew point is below freezing point, it is referred to as the frost point.

CONCEPT OF SATURATION

Saturation is the state of the atmosphere in which air contains the maximum amount of water vapour that it can hold at a specific temperature and air pressure. At saturation the relative humidity is 100%, temperature and dew point are equal, evaporation of water ceases. Unsaturated air can become saturated in three ways.

- evaporation of water into the air from the earth's surface, both land and sea
- the mixing of two masses of air of different temperatures, both initially unsaturated but saturated as a mixture, and
- most commonly, by cooling the air, which reduces its capacity to hold moisture as water vapour sometimes to the point that the water vapour it holds is sufficient for saturation

This atmospheric cooling can be brought about in a number of ways, as by the arrival of a cooler air mass or by the movement of an air mass up a mountain side. If the cooling continues beyond the point of saturation, and provided there are sufficient condensation nuclei in the air around which tiny cloud or fog droplets can form, the excess moisture will condense out of the air as cloud or fog droplets or various forms of precipitation at the earth's surface.

The water vapour content of the atmosphere varies from place to place and from time to time because the humidity capacity of air is determined by temperature. At 30° C, for example, a volume of air can contain up to 4% water vapour. At –40° C, however, it can hold no more than 0.2%. As air temperature increases, more water can remain in a gas phase. As temperature decreases, water molecules slow down and there is a greater chance for them to condense on to surfaces.

Atmospheric water vapour is an important factor in weather for several reasons. It regulates air temperature by absorbing thermal radiation both from the sun and the earth. Moreover, the higher the vapour content of the atmosphere, the more latent energy is available for the generation of storms. In addition, water vapour is the ultimate source of all forms of condensation and precipitation.

PROCESS OF CONDENSATION

Water vapour in the atmosphere is returned to its original liquid state through condensation process. Water vapour can also be converted directly into the solid form through a process called sublimation. Condensation is the direct cause of the various forms of precipitation like dew, frost, fog and clouds. Condensation of water vapour occurs when the temperature of air is lowered to its dew point. It occurs as a result of changes in air volume, temperature, pressure or humidity.

Condensation of moisture occurs in various ways. This is due to the variation in the humidity of the air, in its movement and turbulence, and in its temperature and rate of cooling. Condensation can occur on solid surfaces on the earth as well as in free air above it. The latent heat expended in evaporation is liberated in the process of condensation. During the condensation process, water molecules lose the 600 cal/g of latent heat that was added during the evaporation process. When latent heat is released it is converted into sensible heat which warms the surrounding air. Warming the air increases its buoyancy and fuels the development of storms. Condensation takes place in the presence of condensation nuclei and when the air is nearly saturated. Four mechanisms may lead to condensation.

- When the air is cooled to dew point but its volume remains constant.
- When the volume of the air is increased without addition of heat. In this case, energy is consumed by the work through the adiabatic expansion.
- When a joint change of temperature and volume reduces the moisture holding capacity of the air below its moisture content.
- When evaporation adds moisture to the air.

The most common conditions favouring condensation are those producing a drop in temperature, contact cooling, radiative cooling, mixing of air masses of different temperatures and dynamic cooling of the atmosphere. Contact cooling occurs when warm, moist air passes over a cold land surface. During clear winter nights the surface is cooled due to strong radiation. The surface cooling gradually extends to the moist lower air, reducing its temperature to a point where condensation occurs (the dew point). The mixing of contrasting layers within a single air mass or of two different air masses can also produce condensation. The addition of moisture into the air near the surface by evaporation occurs when cold air moves out over a warm water surface.

Saturation of the air is an essential prerequisite for condensation to occur. Saturation can be brought about either by reducing air temperature or by addition of water vapour into the air. The main ways by which water vapour is added to the air are wind convergence into areas of upward motion, precipitation, evaporation of water from the surface of oceans, waterbodies or wetland, transpiration from plants, cool or dry air movement over warmer water, and lifting of air over mountains. Water vapour needs some kind of surface on which it can condense. For dew and frost, solid objects on the ground provide the required surface. But in the atmosphere, the most needed surfaces for condensation of water vapour are provided by the dust particles or aerosols, which are referred as condensation nuclei.

FORMS OF CONDENSATION

The forms of condensation can be classified on the basis of temperature at which the dew point is reached. For the formation of dew, fog and clouds it is necessary that the dew point is above the freezing point. However, frost and snow are produced when the temperature is lower than the freezing point. Forms of condensation may also be classified on the basis of their location. Dew is formed on the earth's surface, fog and mist are formed near the earth's surface and clouds are formed in the free air.

Fog

Fog is a visible aggregate of minute water droplets suspended in the atmosphere at or near the surface of the earth. In other words, it is a cloud that is in contact with the ground. Fog begins to form when water vapour condenses into tiny liquid water droplets in the air. Since water vapour is colourless, it is actually the small liquid water droplets that are condensed from it that make water suspended in the atmosphere visible in the form of fog.

Fog is considered as a visibility hazard and is defined as microscopically small drops of water condensed from and suspended in air near the surface of the earth in sufficient number to reduce the horizontal visibility to less than one kilometre. Fog normally occurs at a relative humidity near 100%. This can be achieved by either adding moisture to the air or dropping the ambient air temperature. However, fog can also be formed at lower humidity, and sometimes cannot be formed with relative humidity at 100%. Fog formation requires all of the elements that normal cloud formation requires, the most important being condensation nuclei, in the form of dust, aerosols, pollutants, *etc.* for the water to condense upon. When there are exceptional amounts of condensation nuclei present, especially hygroscopic particles such as salt, then the water vapour may condense below 100% relative humidity.

Fog can be formed suddenly, and can be dissipated just as rapidly, depending on what side of the dew point the temperature is on. This phenomenon is known as flash fog. Another common type of formation is associated with sea fog. This is due to the peculiar effect of salt. Clouds of all types require minute hygroscopic particles upon which water vapour can condense. Over the ocean surface, the most common particles are salt from salt spray produced by breaking waves. Condensation on salt particles has been observed to occur at humidity as low as 70%, thus fog can occur even in relatively dry air in suitable locations.

For meteorological purposes, visibility is defined as the farthest horizontal distance at which a person with normal vision can see an object under normal daylight conditions. The reason for degradation of visibility in a foggy atmosphere is the absorption and scattering of natural or artificial illumination by fog particles. The amount of absorption and scattering depends on the microphysical structure of the fog particles, which are called aerosols. Density of fog depends mainly on the number of hygroscopic nuclei. The density of fog is measured on the basis of visibility. Five categories of fogs are recognised on the basis of visibility. In dense fog, buildings or trees situated at a distance of 45 metres are not visible. If the maximum visibility is 180 metre, it is called thick fog. In moderate fog, visibility is 900 metres and in thin fog, it is up to 1.8 km. The term 'mist' is used when the visibility is more than 1 km but less than 2 km, if the relative humidity is 95% or greater. Mist is wet hygroscopic particles suspended in the air. It is less dense fog. It disappears with rising sun. When relative humidity is below 95%, the term 'haze' is used. The blend of fog and smoke, called 'smog', is more common in city and industrial areas. On some occasions toxic materials may be present in fog, smog and haze.

Fog occasionally produces precipitation in the form of drizzle or very light snow. Drizzle occurs when the humidity of fog attains 100% and the minute cloud droplets begin to coalesce into larger droplets. This can occur when the fog layer is lifted and cooled sufficiently, or when it is forcibly compressed from above. Drizzle becomes freezing drizzle when the temperature at the surface drops below the freezing point.

Fog is an important source of moisture for natural vegetation in arid and semiarid areas. It can benefit crops through its contribution in water balance maintenance.

Saturation of moisture occurs either by cooling of air or by evaporation of water into it. Based on these processes fogs can be classified into cooling fogs and evaporation fogs. Cooling fogs are radiation fog, advection fog, upslope fog and inversion fog. Frontal fog and steam fog are evaporation fog types.

Radiation fog: Radiation fog or ground fog is formed by the cooling of land after sunset by thermal radiation in calm conditions with clear sky. The cool ground produces condensation in the nearby air by heat conduction. In perfect calm the fog layer can be less than a metre deep but turbulence can promote a thicker layer. Radiation fogs occur at night, and usually do not last long after sunrise. It is common in autumn and early winter. Radiation fog does not form over oceans because the diurnal variation of temperature is insignificant.

Advection fog: Advection fog occurs when moist air passes over a cool surface by advection (wind) and is cooled. It is most common at sea when tropical air encounters cooler waters, including areas of cold water upwelling. Advection fogs are common over coastal areas by the advection of moist air from sea areas to coastal areas. The advection of fog along the coastline can be propelled onto land by several processes. Sea fogs are developed by the advection of cold continental air to over sea or lake areas. A moderate turbulence is needed for proper development of advection fog.

Upslope fog: Upslope fog or hill fog forms when relatively humid air moves up a slope and adiabatically cools as it rises. If the dew point is reached, the moisture condenses forming an extensive layer of fog. This often causes freezing fog on mountaintops.

Inversion fog: Inversion fog or valley fog forms at the top of a moist layer in mountain valleys, often during winter. It is the result of a temperature inversion caused by heavier cold air settling into a valley, with warmer air passing over the mountains above. It is essentially radiation fog confined by local topography, and can last for several days in calm conditions.

Precipitation fog: Frontal or precipitation fog is formed as precipitation falls into drier air below the cloud and the liquid droplets evaporate into water vapour. The water vapour cools and at the dew point it condenses and fog is formed.

Steam fog: It is formed by intense evaporation of water in cold air. When cool air moves over warm water, enough moisture may evaporate from the water surface to saturate the air immediately above. This type of fog is unstable and observed during autumn in high latitudes in and around water bodies.

Dew

Dew is water in the form of droplets that appear on thin, exposed objects in the morning or evening due to condensation. As the exposed surface cools by radiating its heat, atmospheric moisture condenses at a rate greater than that at which it can evaporate, resulting in the formation of water droplets. The temperature at which water vapour condenses is known as dew point temperature. When the objects on the surface of the earth get cooled in the night at below dew point temperature the water vapour is condensed on these surfaces.

Dew forms best when the atmosphere is clean, the air is calm, content of water vapour in the atmosphere is high, and there is not much wind. With particularly favourable conditions, dew deposition may commence before sunset and continue beyond sunrise. Dew should not be confused with guttation, which is the process by which plants release excess water from the tips of their leaves.

Frost

When temperature is low enough, dew takes the form of ice and this form is called frost. Of course, frost is not frozen dew. Rather, it forms when water vapour changes directly from a gaseous state into solid (ice), without entering the liquid state, a process called sublimation. Condensations for dew and frost formation are practically the same. However, dew forms when condensation occurs on cold objects with the air temperature remaining above the freezing point, whereas frost forms when air temperature falls below freezing point. Frost may be formed due to excessive radiational cooling of ground surface in clear and fairly calm and dry weather. It may also occur due to advection, *i.e.*, wind currents bringing air from cold regions. Because cold air is denser than warm air, in calm weather cold air pools at ground level. This is known as surface temperature inversion. It explains why frost is more common and extensive in low-lying areas Frost is usually translucent in appearance.

Freezing causes injury to plants by extra cellular Ice formation and cellular dehydration. When cell water content is high, ice may form within the cell and kill protoplasm. Frost is always injurious to vegetation. Frost causes economic damage to the farmers when it destroys field crops like castor, tomato, cotton, tobacco, cumin, *etc.* or the fruits in orchards during winter. To reduce the damage due to frost, surface mulching, irrigation, use of large fans to blow away condensing water vapour and large heaters are used. Irrigation during daytime is a common practice to save potato crop from frost.

HUMIDITY AND CROP PRODUCTION

Humidity is one of the fundamental abiotic factors defining any habitat, and is a determinant of the type of animals and plants that can thrive in a given environment. Humidity plays a significant role in weather and climate. It is an important factor in plant growth and crop production as it directly influences the water relations of plant and indirectly affects leaf growth, photosynthesis, pollination, occurrence of diseases and finally economic yield. However, it is not an independent factor but is closely related to rainfall and temperature.

Crop growth

Humidity affects the internal water potential of plants. It is a major determinant of potential evapotranspiration. So, it determines the water requirement of crops. Humidity also influences certain physiological phenomena including transpiration. Leaf growth not only depends on biochemical processes but also upon the physical process of cell enlargement. Cell enlargement occurs as a result of turgor pressure developed within the cells. Turgor pressure is higher under high relative humidity due to less transpiration. Thus leaf enlargement is higher in humid areas.

When humidity is low, moisture evaporates from plants very quickly. When this happens for a longer period, plants wilt rapidly as too much water is pulled out of plant cells through transpiration. Conversely, when humidity and temperature are both high, plants can get overheated because of reduced transpiration, thus restricting evaporative cooling.

Change in relative humidity can produce various morphological and anatomical changes in the plants. For example, orchids grow abundantly in humid forests as epiphytes depend for their moisture requirement on the atmosphere by developing certain morphological and anatomical characteristics that are not found in other plants (hydroscopic aerial roots). In desert region where relative humidity is low, xerophytes show certain adaptations to conserve water. High relative humidity can prolong the survival of crops under moisture stress.

Photosynthesis

Photosynthesis is indirectly affected by relative humidity. When relative humidity is low, transpiration increases causing water deficits in the plant. Water deficits cause partial or full closure of stomata and increase mesophyll resistance blocking entry of carbon dioxide. This leads to reduced photosynthesis.

Pollination

Moderately low air humidity is favourable for seed set in many crops, provided soil moisture supply is adequate. At high relative humidity pollen may not be dispersed from the anthers. Humidity can also affect the fruit set of some plant species. An example is the bean (*Phaseolus* spp.), which has been shown to respond negatively to low humidity during fruit set.

Disease and pest

Relative humidity plays a significant role in the outbreak of disease and pest epidemics. The incidence of insect pests and diseases is higher under high humidity conditions. High humidity promotes the growth of some saprophytic and parasitic fungi and bacteria which cause various plant diseases. High relative humidity favours easy germination of fungal spores on plant leaves. For example, the blight diseases of potato and tea spread more rapidly under humid conditions. Several insects such as aphids and jassids thrive better under moist conditions.

Storage of grains and fruits

Humidity is also an important consideration for post-harvest storage of crops. Cold temperature and low humidity are favourable for the long-term storage of grains (rice, wheat, maize, pulses, *etc.*). Conversely, fleshy vegetables and fruits (apples, pineapples, cucumbers, brinjal, tomato, okra, *etc.*) need to be stored in cold temperatures and high humidity in order to prevent water loss from the fruit that results in shrivelling.

Grain yield

Very high or very low relative humidity is not conducive for higher grain yields. Moderately high relative humidity of 60–70% is beneficial for most of the commercial crops. Reduced transpiration due to very high humidity adversely influences translocation of plant nutrients. Under very high relative humidity, grain yields of maize and wheat are reduced. This reduction could be attributed to adverse effect of relative humidity on pollination and high incidence of pests. Increase in relative humidity during panicle initiation to maturity increases grain yield of sorghum under low humidity conditions due to favourable influence of relative humidity on water relations of plants and photosynthesis. With similar amount of solar radiation, crops that are grown with irrigation give lower yields compared to those grown with equal amount of water as rainfall. This is because the dry atmosphere, which is little affected by irrigation, independently suppresses the growth of crops.

IMPORTANCE OF DEW IN AGRICULTURE

Due to its dependence on radiation balance, dew amounts can reach a theoretical maximum of about 0.8 mm per night, measured values, however, rarely exceeding 0.5 mm. In most climates of the world, the annual average is too small to compete with rain. Under suitable conditions, in semiarid areas, it may exceed even 25–30 mm per annum. In parts of Israel, dew may supply as much as 55 mm of water annually.

Dew is an important secondary source of moisture for crops during the non-rainy season and plays vital role in plant growth. Dew occurrences benefit the plants in many ways. The dew deposited on the leaf surface in the morning delays the rise in leaf temperature and thus reduces the rate of transpiration and minimises crop water requirements. Dew provides water for direct plant use. Foliar absorption of dew is an important factor in survival of natural vegetation in arid regions. In regions with considerable dry seasons, adapted plants like lichen or pine seedlings benefit from dew. Further, this moisture is available mainly during the dry summer months when plants are experiencing the greatest moisture stress.

Increased humidity due to dew formation in the crop canopy may accelerate pest and disease build-up. Dew encourages development of habitats for pathogens such as the fungus *Phytophthora infestans* which infects potato plants.

Chapter 8

Cloud

Clouds are visible mass of little drops of water or frozen crystals suspended in the atmosphere above the surface of the earth. The World Meteorological Organisation defined cloud as a visible aggregate of minute particles of water or ice or both, in the free air. This aggregate may include larger particles of water or ice and particles such as those present in fumes, smoke or dust.

Clouds are the most important form of suspended water droplets caused by condensation. The typical size of these water droplets and ice crystals range from less than 1 μm to 50 μm. When surrounded by billions of other droplets or crystals, they become visible as clouds. In the atmosphere sufficient moisture is required to allow condensation to take place. Then, moisture needs a suitable surface upon which it can condense. In free air, condensation begins on hygroscopic nuclei such as aerosols. These are microscopic particles and include dust, smoke, salts and chemical compounds. Nuclei range in size from 0.001 μm radius to over 10 μm. On an average, oceanic air contains one million condensation nuclei per litre and land air holds some 5 or 6 millions. On hygroscopic particles such as sea salts, condensation can begin before the air is saturated. Supersaturation in clouds rarely exceeds one per cent. On average, clouds contain only four per cent of the total water in the atmosphere at any one time. Almost all clouds exist in the troposphere, although the tops of some severe thunderstorms occasionally extend to the stratosphere. Clouds move along with the wind and remain stationary in calm winds. They grow vertically and horizontally, depending on the temperature of the cloud and its ambient temperature. They may dissipate or precipitate. Under favourable atmospheric conditions, they burst and fall as rain. Clouds are essential to the earth-atmosphere system. They complete the following functions.

1. Clouds help regulate earth's energy balance by reflecting and scattering solar radiation and by absorbing earth's infrared energy.
2. Clouds are required for precipitation to occur and, hence are an essential part of the hydrologic cycle.

3. Clouds indicate what type of atmospheric processes are occurring (*e.g.*, cumulus clouds indicate surface heating and atmospheric turbulence).
4. Clouds help redistribute extra heat from the equator towards the poles.
5. Clouds are the visible expression of the atmospheric stability. Watching the variety of cloudscapes helps to understand how the weather is changing.

FORMATION OF CLOUDS

The formation of clouds depends on atmospheric stability and vertical motion but it also involves micro scale processes. Clouds are formed when air containing water vapour is cooled below a critical temperature called the dew point and the resulting moisture condenses into droplets on microscopic dust particles known as condensation nuclei in the atmosphere. There are three basic requirements for clouds to occur.

1. Water vapour must be present in sufficient amounts so that saturation can be reached by some means.
2. Cloud condensation nuclei (CCN) must be present to provide a surface on which water will condense. Examples of CCN include dust in the air from the earth's surface, salt particles from the sea, combustion products, and volcanic or meteorite dust.
3. Cooling mechanism is required to cool the air temperature to the dew point temperature.

The requirements for sufficient water vapour and CCN are rarely the limiting factors for cloud development. Typically, the limiting factor is a cooling mechanism. Thus, the air temperature needs to be lowered to the dew point temperature for a cloud to form. If a cooling mechanism is not present, clouds will not form. The air is normally cooled by expansion during its upward movement.

Mechanism of air uplift

Orographic uplift, frontal uplift, convergence and localised convection are the four mechanisms by which air is lifted leading to its adiabatic cooling and hence condensation and cloud formation.

Orographic uplift: In orographic uplift air flowing towards a hill or mountain is deflected around and above the barrier. The upward displacement of air leads to adiabatic cooling and possible cloud formation. The thickness or height of clouds generally greatly exceeds the height of the barrier that helped create it.

Frontal uplift: Frontal uplift occurs at the meeting place of two different air masses. When cold air advances towards warmer air (cold front situation) the denser cold air displaces the lighter warm air ahead of it. When warm air flows towards cold air (a warm front), the warm air is forced upward.

Convergence: When a low pressure zone is developed near the surface, it causes the convergence of the atmosphere at the centre of low pressure with air flowing from all directions. It causes vertical motion of the air near the centre.

Localised convection: Heating of the earth's surface can produce free localised convection over a fairly limited area. Free convection arises from buoyancy, the tendency for a lighter fluid to float upward through a denser one. Buoyancy can initiate uplift as well as can speed up or slow down the uplift begun by orographic effect, frontal lifting and convergence.

Atmospheric stability

Atmospheric stability refers to the air's susceptibility to uplift. Atmospheric stability is closely related to buoyancy which is determined by the density (hence temperature) differences between an air parcel and its surroundings. Temperature in the air parcel is governed by either dry or saturated adiabatic lapse rates whereas that in the surroundings is governed by the environmental lapse rate. Thus, atmospheric stability can be determined by comparing the environmental lapse rate to the saturated and dry adiabatic lapse rates. Three types of static stability are recognised. They are absolutely stable, absolutely unstable and conditionally unstable.

Absolutely stable: If the environmental lapse rate is less than the saturated or moist adiabatic lapse rate, the air is absolutely stable *i.e.*, rising air will cool faster than the surrounding air and lose buoyancy. The stable air resists upward displacement and sinks back to its original level when the lifting mechanism ceases, regardless of whether it is saturated or unsaturated. This often happens in the early morning, when the air near the ground has cooled overnight. Cloud formation in stable air is unlikely. When the temperature remains constant or increases with height, the air will be absolutely stable.

Absolutely unstable: Whenever the environmental lapse rate exceeds the dry adiabatic lapse rate, the air is absolutely unstable. An unstable air parcel becomes buoyant when lifted and continues to rise upward, even after the original lifting process ceases, regardless of whether it is saturated or not. This often happens in the afternoon over many land masses. In these conditions, the likelihood of cumulus clouds, showers or even thunderstorms is increased.

Conditionally unstable: If the environmental lapse rate is between the moist and dry adiabatic lapse rates, the air is conditionally unstable. The tendency for a lifted parcel to sink or continue to rise depends on whether or not it becomes saturated and how far it is lifted. The parcel becomes buoyant only if it is forced upward some distance above the level at which saturation occurs (known as the lifting condensation level, LCL). The height to which a parcel of conditionally unstable air must be lifted for it to become buoyant and to rise on its own is called level of free convection (LFC). When the air rises above that level, it is common for clouds to rapidly increase in depth and yield precipitation.

All these three types of static stability describe what would happen to a small parcel of air that is lifted or lowered while the surrounding air is kept in place. On the other hand potential or convective stability describes what happens when entire layers of air are displaced upward. With such upward movement, the environmental lapse rate may be altered so that the statically stable conditions within the layer can change to unstable conditions or *vice versa.* Potential instability is particularly important for forecasting of severe thunderstorms. Stability can define whether or not clouds will form, where they will form, when they will form and what type of clouds they will be.

CLASSIFICATION OF CLOUDS

Luke Howard, an English amateur meteorologist with a strong grounding in the Latin language, developed the basis for tropospheric cloud classifications in 1802. He believed that the changing cloud forms in the sky could unlock the key to weather forecasting. Howard used universally accepted Latin language for the nomenclature, which caught on quickly. It was formally adopted by the International Meteorological Commission in 1929.

Because of the large range in temperatures and air movement in the troposphere, clouds vary in structure and the variety of cloud types seems to be baffling. However, clouds can be classified into different types according to their origin and nature. Cloud names devised by Howard remain to this day as the standard international system for cloud identification. The first international cloud atlas (International Atlas of Clouds and of States of the Sky) by International Meteorological Congress was prepared in 1896 with subsequent revisions in 1932 and 1939. The World Meteorological Organisation (WMO) prepared the relatively recent world cloud atlas in 1956.

The internationally adopted system of cloud classification is based on the general shape, structure and vertical extent of the clouds and their altitude. The clouds are classified into 4 cloud families, which are in different heights of the troposphere.

1. Family A- high level clouds, altitude of 5–13 km
2. Family B- medium level clouds, altitude of 2–7 km
3. Family C- low level clouds, altitude of 0–2 km
4. Family D- clouds with large vertical, altitude from ground surface extending up to 13 km

Cloud families are named by the base height, not the cloud top, and bases may differ depending on the geographical zone. Besides, three basic cloud forms are recognised *i.e.*, Cirrus, Cumulus and Stratus.

1. Cirrus clouds are high, white and thin displaying delicate streaks or lines. They are separated or detached clouds appearing in veil like patches and often have a feathery appearance. In Latin *Cirrus* means curl or filament.
2. Cumulus clouds consist of globular individual cloud masses. Normally they exhibit flat base and appear as rising domes or towers. Such clouds are frequently described as having a cauliflower like structure.
3. Stratus clouds are best described as sheets or layers (strata) that cover much of the sky.

Although there may be minor breaks, there are no distinct individual cloud units. All clouds have one of these three basic forms or combinations or modifications of them. Ten basic cloud types called cloud genera are recognised with the combination of the height and form criteria. Cloud types are universal yet each cloud is unique. Most types of clouds can be found in all climate zones throughout the world, though the frequency and size of particular cloud types do vary. Definite weather patterns may be associated with specific clouds.

High level clouds (Family A)

These clouds form between 3 and 8 km in the polar regions, 5 and 12 km in the temperate regions, and 6 and 18 km in the tropical regions. High level clouds are given the prefix 'cirro'. Three cloud types make up the family of high clouds, cirrus (Ci), cirrocumulus (Cc) and cirrostratus (Cs). Because of low temperature and small amount of water vapour present at high altitudes, all high clouds are thin and white and made up of primarily ice crystals. High clouds generally are not precipitation makers. However, when cirrus clouds give way to cirrocumulus clouds, a stormy weather is predicted.

Cirrus: Cirrus is a detached cloud, composed of white, delicate, ice filaments. Winds aloft often cause these fibrous ice trails to bend or curl.

Cirrostratus: Cirrostratus is a transparent, white cloud veil of fibrous or smooth appearances that may cover much or all of the sky. Sometimes it produces haloes around the sun or moon.

Cirrocumulus: Cirrocumulus clouds appear as white patches composed of very small cells or ripples. Often the small globules of this cloud are arranged in a regular pattern. This pattern is called mackerel sky due to its similarity to the pattern formed by fish scales.

Middle level clouds (Family B)

These clouds tend to form at about 2 km but may form at heights up to 8 km. Depending on the altitude, time of year, and vertical temperature structure of the troposphere, these clouds may be composed of liquid water droplets, ice crystals, or a combination of the two, including supercooled droplets (*i.e.*, liquid droplets whose temperatures are below freezing). These clouds are brighter and less fragmented in appearance due to their distance from the ground and the higher composition of ice crystals. Middle level clouds vary in thickness from relatively flat sheets of cloud to a more cumuliform appearance. These clouds tend to have apparent speeds slower than the lower level clouds. Middle level clouds are given the prefix 'alto'. There are 2 basic types of middle level clouds, altocumulus (Ac) and altostratus (As).

Altocumulus: Altocumulus clouds generally have cumuliform tops and bases that are usually relatively darker than the tops. They can be widespread or patchy depending on the conditions. They vary in appearance from broken to smooth, and vary in thickness. Altocumulus clouds indicate unstable or unsettled conditions. They can develop from dissipating thunderstorms during the morning and then redevelop during the day if the air remains unstable. Altocumulus varies in its apparent movement (speed) depending on the wind and direction at that level. They develop in more than one layer and also in conjunction with other cloud types. If altocumulus develops into thicker layers, precipitation can develop. The expected intensity of rainfall from altocumulus is most often light to moderate.

Altostratus: Altostratus refers to middle level cloud that appears as a flat, smooth dark grey sheet. They can develop into a thick or thin layer. As a thin layer, the sun can be observed through the cloud. In its thicker form, the sun can only occasionally be observed through the thinner sections. Obviously, the thicker the altostratus, the darker it becomes. When observed closely, it becomes apparent that altostratus is not just one complete layer but a composition of several thin layers. Altostratus commonly develops during afternoon or evenings particularly when a trough of low pressure or cold front approaches.

Low level clouds (Family C)

These clouds consist of clouds in the lower layers of the atmosphere up to 2 km. Being closer to the ground, lower level clouds appear to move or progress faster than other clouds. The clouds generally move in the direction very similar to the direction of the wind on the ground. When observed closely, it is easy to observe its ever-changing structure. Low-level clouds are not given a prefix, although their names are derived from 'strato' or 'cumulo', depending on their characteristics. There are 3 main types of lower level clouds, nimbostratus (Ns), stratocumulus (Sc) and stratus (St).

Nimbostratus: Nimbostratus clouds derive their name from Latin *nimbus* meaning rain clouds and *stratus* meaning to cover with a layer. These clouds are one of the chief precipitation producers. They form in association with stable conditions. Forced ascent of stable air leads to the formation of stratified cloud layers with large horizontal extent compared to thickness. The precipitation associated with nimbostratus clouds is generally light to moderate but of long duration and widespread.

Stratocumulus: Stratocumulus is a low cloud that generally moves faster than cumulus and is not well defined in appearance. They tend to spread more horizontally rather than vertically. The bases of stratocumulus are normally darker than the tops. However, they can vary in terms of characteristics. They may appear as large flat areas of low, grey cloud. Sometimes they appear in the form of rolling patches of cloud aligned parallel to each other. Stratocumulus can also appear in the form of broken clouds or globules. The sun, moon and generally the sky can be observed through the breaks in broken stratocumulus clouds. The amount of stratocumulus covering the sky depends on the amount of moisture concentrated at that level of the atmosphere. The speed at which the cloud moves varies according to the wind speed at that level.

Stratus: Stratus is defined as low cloud that appears fragmented and thin. It can also occur in the form of a layer or sheet. Stratus lacks the vertical growth of cumulus and stratocumulus, and therefore lacks the contrast. Being closest to the ground, stratus clouds normally move fairly rapidly in the direction of the wind depending on the wind speed. Like stratocumulus, stratus develops under several conditions or weather situations. Stratus mostly develops under the influence of wind streams where moisture condenses in the lower layers of the atmosphere. Changes in wind during the summer months often lead to the development of stratus as the wind evaporates moisture from the ocean and mixes the surface air with the cooler air above. In these conditions, stratus develops in patches and gradually may become widespread forming stratocumulus. Stratus, like stratocumulus, can develop in weather conditions associated with thunderstorms. With light winds, stratus is normally observed in sheets. In stronger wind conditions, stratus develops in patches, similar in appearance to stratocumulus. The stratus is the best example of a cloud that displays rapid development and dissipation. Both the direction and appearance of stratus can change rapidly with changing weather conditions. It can be clear and redevelop several times during certain conditions usually appearing when rain approaches, and clearing as the rain recedes. Being the lowest cloud layer, it obscures at least partially the view of stratocumulus or other types of clouds above.

Vertical clouds (Family D)

These clouds can have strong up-currents, rise far above their bases and form at many heights. Vertical clouds include cumulus (Cu) and cumulonimbus (Cb).

Cumulus: Cumulus is a cauliflower-shaped low level cloud with dark bases and bright tops. The bases of cumulus are usually flat. Cumulus often occurs in conjunction with other clouds and may vary in appearance. If cumulus is observed below other clouds, the shadow effect of other clouds can decrease contrast of the cumulus. Cumulus is observed to develop under various types of conditions. The most common conditions are that of light winds where cumulus remains almost stationary. Quite often under these conditions, cumulus develops over hills or mountains. They are generally not fast moving clouds. Depending on the height of the atmosphere where cumulus develops, it will move in the direction of the wind at that level. At lower levels, cumulus will tend to move in a direction similar to the direction of the wind at the surface. At higher levels, particularly on days when cumulus may develop into thunderstorms, it may move in a different direction than that of the surface winds. Cumulus clouds are generally associated with fair weather.

Cumulonimbus: Cumulonimbus is a dark, dense, billowy cloud of considerable vertical extent in the form of huge towers. The top of these clouds frequently spread out in the shape of an anvil. They are associated with heavy precipitation and thunderstorms.

The summary of ten principal cloud types is presented in table 8.1 and different cloud plates are given in Fig 8.1.

Table 8.1 Summary of ten principal cloud types

Name	Abbreviation	Description
Cirrus	Ci	Fibrous, threadlike, white feather clouds of ice crystals, whose form resembles hair curls.
Cirrostratus	Cs	Milky, translucent cloud veil of ice crystals, which sometimes causes halo appearances around moon and sun.
Cirrocumulus	Cc	Fleecy cloud; Cloud banks of small, white flakes.
Altocumulus	Ac	Grey cloud bundles, sheds or rollers, compound like rough fleecy cloud, which are often arranged in banks.
Altostratus	As	Dense, grey layer cloud, often evenly and opaquely, which lets the sun shine through only a little.
Nimbostratus	Ns	Rain cloud. Grey, dark layer cloud, indistinct outlines.
Stratocumulus	Sc	Cloud plaices, rollers or banks compound dark grey layer cloud.
Stratus	St	Evenly grey, low layer cloud, which causes fog or fine precipitation and is sometimes frazzled.
Cumulus	Cu	Heap cloud with flat basis in the middle or lower level, whose vertical development reminds of the form of towers, cauliflower or cotton.
Cumulonimbus	Cb	In the middle or lower level developing thunder-cloud, which mostly up-rises into the upper level.

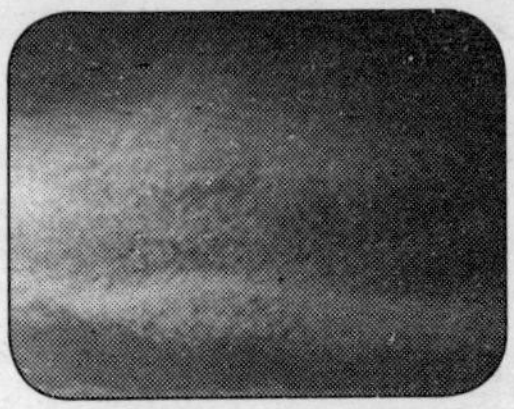
Cirrus

Cirrostratus

Cirrocumulus

Family- A

Altocumulus

Altostratus

Family-B

Nimbostratus Stratocumulus Stratus

Family-C

Cumulus Cumulonimbus

Family-D

Fig. 8.1 Cloud genera

(*Source*: http://www.clouds-online.com)

Under each cloud genus there are certain species, varieties and special forms, which describe the appearance of the cloud genera nearer. The 14 species are described based on the shape and structure of the clouds. They are *calvus, capillatus, castellanus, congestus, fibratus, floccus, fractus, humilis, lenticularis, mediocris, nebulosus, spissatus, stratiformis* and *uncinus.* Similarly, there are 9 varieties based on the transparency and arrangement of macroscopic elements. Clouds may develop in clear air or from other cloud, called mother clouds. When it is developed from another cloud, the designation of the mother cloud is attached with a 'genitus'. Thus if a cumulonimbus develops from a cumulus, this cloud is designated as cumulonimbus cumulogenitus (Cb cugen). If the new cloud family did not develop yet, but the development goes there, the designation of the current cloud family is attached with a 'mutatus'. Thus if the development of a stratus goes to a stratocumulus, this cloud is designated as stratus stratocumulomutatus (st scmut).

Other interesting clouds

Wall cloud: A localised cloud lowering from the rain-free base of a strong thunderstorm. The lowering denotes a storm's updraft where rapidly rising air causes lower pressure just below the main updraft, which enhances condensation and cloud formation just under the primary cloud base. Wall clouds take on many shapes and sizes. Some exhibit strong upward motion and cyclonic rotation, leading to tornado formation, while others do not rotate and essentially are harmless.

Shelf cloud: A low, horizontal, sometimes wedge-shaped cloud associated with the leading edge of a thunderstorm outflow or gust front and potentially strong winds.

Fractus: Low, ragged stratiform or cumuliform cloud elements that normally are unattached to larger thunderstorm or cold frontal cloud bases. Also known as scud, fractus clouds can look ominous, but by themselves are not dangerous.

Contrail: Narrow, elongated cloud formed as jet aircraft exhaust condenses in cold air at high altitudes, indicative of upper level humidity and wind drift.

ARTIFICIAL RAINMAKING

Rainmaking, also known as artificial precipitation is a practice of artificially inducing or increasing precipitation through clouds by adding external agents. This process is known as cloud seeding. This stimulation is done by aeroplanes or rockets.

Cloud seeding is the process of spreading either dry ice (solid carbon dioxide), or more commonly, silver iodide into the upper part of clouds to try to stimulate the precipitation process and form rain. Since most rainfall starts through the growth of ice crystals from supercooled cloud droplets in the upper parts of clouds, the silver iodide particles are meant to encourage the growth of new ice particles. There are three cloud seeding methods, *viz.*, static, dynamic and hygroscopic.

Static cloud seeding: Static cloud seeding involves spreading of a chemical like silver iodide into clouds. The silver iodide provides a crystal around which moisture can condense. The moisture is already present in the clouds, but silver iodide essentially makes rain clouds more effective at dispensing their water.

Dynamic cloud seeding: Dynamic cloud seeding aims at boosting vertical air currents, which encourages more water to pass through the clouds, translating into more rain. Up to 100 times more ice crystals are used in dynamic cloud seeding than in the static method. The process is considered more complex than static clouding seeding because it depends on a sequence of events working properly. An unexpected outcome in one stage could ruin the entire process, making the technique less dependable than static cloud seeding.

Hygroscopic cloud seeding: Hygroscopic cloud seeding disperses salts through flares or explosives in the lower portions of clouds. The salts grow in size as water joins with them.

The most common chemicals used for cloud seeding include silver iodide, potassium iodide and dry ice (solid carbon dioxide). Liquid propane, which expands into a gas, has also been used. This can produce ice crystals at higher temperatures than silver iodide. The use of hygroscopic materials, such as common salt, is becoming more popular for cloud seeding.

Cloud seeding can be used to enhance precipitation and has been used to prevent areas from destruction from drought. However, exploitation of rain in one region through cloud seeding may upset the hydrological balance over the other regions. Although it is considered a mainstream scientific tool to improve rain precipitation and snow, the history of cloud seeding has experienced uncertain results because it can never be known whether a cloud that rains after seeding might have rained anyway. This is because seeding is performed on clouds that look like they have some potential for producing rain.

There is evidence that clouds, especially in the weather-making troposphere, contain biological ice nuclei that may play a key role in the formation of precipitation. Bioprecipitation, the concept of rain-making bacteria, was proposed by David Sands. Such microbes, called ice nucleators are found in rain, snow, and hail throughout the world.

CLOUD DISSIPATION

Cloud dissipation is a process opposite to that of cloud formation. Clouds are composed of liquid suspended water droplets in about a 100% relative humidity environment. The clouds dissipate by three primary ways; increase in temperature, cloud mixing with drier air and air sinking within the cloud. Often, more than one of these processes occur simultaneously to erode a cloud.

1. When the temperature increases, the air has a higher capacity to evaporate liquid water. As a cloud's temperature increases, evaporation occurs and it reduces the liquid moisture content of the cloud. A cloud can be warmed by solar radiation and longwave emission from the carth's surface. Daytime heating increases the capacity of the air to evaporate liquid water. Low clouds, as such are often dissipated due to daytime heating. The power of daytime heating to erode clouds depends on the season, the cloud thickness and the overall stability and the lift present in the troposphere.
2. A cloud does not remain perfectly adiabatic. Some environmental air does mix into the cloud mass. If a cloud is no longer developing and not adding additional condensational moisture, the drier environmental air will gradually erode the cloud. Instead of having a sharp defined appearance, after mixing with environmental air the cloud will look wispy with edges that are not well defined. This process is called entrainment. During entrainment, drier air incorporates itself into the cloud and induces evaporation.
3. When air sinks, it warms adiabatically. Again, warming will induce evaporation and erosion of the cloud. This can occur when dynamic sinking mechanisms instigate or increase over a cloud or a cloud field.

Chapter 9

Precipitation

Precipitation is water in liquid or solid forms, falling on to the earth. In other words, precipitation is the deposition of atmospheric moisture on earth. Formation of clouds does not mean precipitation. Precipitation involves condensation, growth of hygroscopic nuclei, cooling of nuclei at different stages and their fall on to the earth's surface. There may be several clouds but no precipitation. If the cloud particles are too small and remain suspended in the sky, they may not fall on the earth. Under certain conditions, water particles formed tend to fall but do not reach the surface since they get evaporated on the way. Only when the cloud droplets or ice crystals grow to certain size and can no more remain floating due to buoyancy of air, do they fall as precipitation.

Precipitation is the major source of water to all the living organisms on the earth. All the water sources on the earth depend on precipitation only.

FORMS OF PRECIPITATION

Precipitation form is the state that the moisture is in, *i.e.*, liquid, freezing, or frozen (solid). Different forms of precipitation may occur together, such as mixed rain and snow; but such an occurrence is simply a mixture of forms, not a separate form of precipitation. Because atmospheric conditions vary greatly both geographically and seasonally, several different forms of precipitation are possible. All forms of precipitation are collectively termed hydrometeors. Hydrometeors have been classified into 50 specific types. Out of these precipitation forms, only rain (liquid form) and snow (solid form) make significant contribution to total precipitation. In many parts of the world, the term rainfall is used interchangeably with precipitation.

Liquid forms

Liquid precipitation is any precipitation that falls as a liquid and remains liquid after striking an object, such as the earth's surface. Rain, drizzle and mist are the forms of liquid precipitation.

1. The term rain is restricted to drops of water that fall from a cloud and have a diameter of at least 0.5 mm. Their typical size is 2 mm. No matter what the rainfall intensity is, the size of raindrops rarely exceed 5 mm. Large drops cannot survive due to increasing surface tension and they regularly break into smaller ones. Rainfall has an accumulation rate of more than 1 mm/hour. Most of the rainfall begins as snow crystals and melt after entering the warmer air below the cloud. They reach the ground as raindrops.
2. Fine uniform drops of water having a diameter less than 0.5 mm are called drizzle. Usually drizzle falls from low level stratus clouds and is frequently accompanied by fog. Size of drizzle droplets is larger than the droplets in clouds, but smaller than raindrops.
3. Precipitation containing the very small droplets able to reach the ground is called mist. Mist can be so fine that the tiny droplets appear to float.

Freezing forms

Freezing precipitation is any precipitation that falls as a liquid and freezes upon contact with an object, such as freezing rain or freezing drizzle. In this form of precipitation, the liquid water may be a super-cooled liquid and freeze upon contact with an object, or the water droplet may have an above freezing temperature and freeze upon contact with an object that has a temperature below freezing. Supercooling occurs in clouds where droplets remain in a liquid form in temperatures below the normal freezing point. Rain or drizzle which freezes on contact within a subfreezing air mass gains the preceding word of freezing, becoming known as freezing rain or freezing drizzle. Freezing rain and freezing drizzle are also known as glaze and sleet, respectively.

Frozen forms

Frozen precipitation is any precipitation of water that falls in its solid state, such as snow, hail, ice pellets, *etc*.

Snow: Snow is defined as precipitation of water in the solid form of small or large ice crystals. It occurs only when the condensing medium has a temperature below freezing temperature. Snow is generally in the form of individual crystals or in flakes that are aggregates of many crystals, called snowflakes. Snowflakes are formed in high clouds. Wet snow is crystals bonded by liquid water in the interior pores and crevices. Individual crystals have hexagonal forms. The crystal structure of snow has very fine and needlelike branches. Snow is measured with snow gauge.

Snow grain: Snow crystals less than 1 mm in diameter are called snow grains. These are white, opaque grains of ice similar in crystal structure to snow.

The crystal structure of snow grains has thicker, denser elements, with the space between hexagonal branches being commonly completely filled. Snow grains do not bounce or shatter on hard surfaces. They usually fall in small quantities, mostly from stratus clouds and never as showers.

Diamond dust: Diamond dust, sometimes just called ice crystals, consists of extremely small ice crystals, usually formed at low levels at temperatures below –30° C. The name diamond dust comes from the sparkling effect created when light reflects on the ice crystals in the air.

Hail: Hail is a form of solid precipitation. It consists of balls or irregular lumps of ice, each of which is referred to as a hailstone. Hailstones on earth consist mostly of water ice and measure between 5 mm and 200 mm in diameter, with the larger stones coming from severe thunderstorms. Hail frequently displays a layered appearance of alternate opaque and clear ice. It is produced only in thunderstorms, but may be ejected from the top or sides of a thunderstorm to fall and strike the ground without a cumulonimbus cloud directly overhead. Any thunderstorm which produces hail that reaches the ground is known as a hailstorm.

Graupel: Graupel or snow pellets or small hail are white, opaque, round or occasionally conical shaped kernels of snowlike consistency, 2 to 5 mm in diameter. These are crisp, easily compressible, and may rebound or burst upon striking a hard surface. Snow pellets occur almost exclusively in snow showers.

Rime: Rime is the clear crystalline or granular ice deposited when supercooled fog or cloud droplets encounter a vertical structure, trees or suspended cables. The rime deposits grow into the wind in a triangular form related to the wind speed. It is common in cold, maritime climates and on mid-latitude mountains in winter.

Some important forms of precipitation are summarised in the table 9.1.

Table 9.1 Some important forms of precipitation

Type	Approximate size (diameter in mm)	State of water	Description
Mist	0.005–0.05	Liquid	Droplets large enough to be felt on the face when air is moving @ 1m/s, associated with stratus cloud
Drizzle	0.05–0.5	Liquid	Small uniform drops that fall from stratus clouds, generally for several hours
Rain	0.5–5	Liquid	Generally produced by nimbostratus or cumulonimbus clouds

Type	Approximate size (diameter in mm)	State of water	Description
Sleet	0.5–5	Solid	Small spherical to lumpy ice particles that form when raindrops freeze while falling through a subfreezing air layer
Glaze	1 mm to 20 mm thick layer	Solid	Produced when supercooled raindrops freeze on contact with solid objects
Rime	Variable accumulation	Solid	Frostlike deposits usually consisting of ice features that point into the wind
Snow	1–2	Solid	Crystalline snows can assume many shapes including six sided crystals, plates and needles
Hail	5–100	Solid	Hard rounded pellets of ice produced in large convective cumulonimbus clouds where frozen ice particles and super-cooled water coexist
Graupel	2–5	Solid	Irregular mass of soft ice formed when rime collects on snow crystals

PROCESSES OF PRECIPITATION

The main ways by which water vapour is added to the air are wind convergence into areas of upward motion, precipitation or virga falling from above (in certain conditions precipitation may fall from a cloud but then evaporates or sublimes before reaching the ground; this is termed as 'virga' and is more often seen in hot and dry climates), evaporating water from the surface of oceans, water bodies or wet land, transpiration from plants, cool or dry air moving over warmer water, and lifting air over mountains. Water vapour normally begins to condense on condensation nuclei such as dust, ice, and salt in order to form clouds.

Precipitation processes involve four steps. They are development of condensation nuclei, initiation of condensation, growth of condensed product and finally precipitation. Thus, the first step of precipitation is condensation when water vapour changes to liquid water droplets or clouds. However, not all condensation is followed by precipitation. Only when cloud droplets or ice crystals grow to such a large size as to overcome the normal buoyancy and updrafts in the atmosphere, then precipitation can occur. While the size of cloud droplets seldom exceeds 10 micron (μm), size of raindrops vary from 200 to 5000 μm. Thus, a raindrop is large enough to reach the ground without getting evaporated.

The exact process that causes the cloud droplets to join together to form large raindrops capable of falling on the earth as precipitation are not fully understood. Many earlier theories of raindrop met with objections. However, two mechanisms have been postulated to explain these processes. The rapid growth of raindrops involves either the growth of ice crystals at the expense of water drops (Bergeron-Findeisen theory) or the coalescence of small droplets by the sweeping action of the falling drops. A cloud can release precipitation only when there is a colloidal instability in it. Hence, any process that brings about an element of colloidal instability in clouds may be said to be responsible for precipitation.

Bergeron-Findeisen's ice crystal theory

This theory was proposed by Tor Bergeron, a meteorologist from Sweden in 1933 and further developed by another meteorologist Walter Findeisen from Germany. Water droplets in the cloud do not freeze at 0° C. When water remains in liquid form below 0° C, it is referred to as supercooled. In the atmosphere, supercooled water has been observed down to –40° C. The supercooled water tends to freeze when it is disturbed or come in contact with the freezing nuclei. Freezing nuclei are ice crystals of the size one microns and are sparse in the atmosphere. When the ascending air rises well above the freezing level, some water droplets will be changed into ice and through sublimation water vapour will enter into solid state. If a single ice-crystal is introduced into a cloud of supercooled water droplets, the entire cloud rapidly changes over to an all-ice cloud. In subzero temperature, the saturation vapour pressure of ice is lower than that of the water. That means, when air is saturated with respect to water, it is oversaturated with respect to ice. Thus, vapour is diffused rapidly from air to ice crystals. The growth of ice crystals is rapid enough to generate crystals large enough to fall. While falling from the clouds, the ice crystals grow by intercepting cloud droplets that freeze upon them. Sometimes these falling ice crystals are broken up into fragments which again become freezing nuclei for other droplets. A chain reaction, thus takes place. These ice crystals grow further by accretion process and become snowflakes before leaving the clouds. Snowflakes generally melt before reaching the ground and fall as rain. This process is believed to be the dominant process in mid and high latitudes, where clouds ascend to an altitude of subzero temperature. Artificial rain making is based on these mechanisms.

Collision-coalescence theory

This theory was developed by E. G. Boven of Australia in 1940s. Ice crystals contribute little in the development of precipitation from the warm clouds in tropical region, where temperature does not go below 7° C. The occurrence of precipitation from such clouds involves the coalescence of cloud droplets of different sizes. Since the rate of fall of these unequal particles is different, they collide with each other within the cloud, and the larger drops grow at the expense of the smaller ones. The rate of growth of falling water droplets depends on variables such as the size and distribution of drops and their concentration in the cloud. Clouds made up of exclusively liquid droplets can generate precipitation only when they contain droplets larger than 20 microns. The hygroscopic materials such as sea salt can provide the giant condensation nuclei on which large water drops can develop. Larger drops fall faster and collide with the smaller droplets and coalesce. In a cloud with great vertical thickness and abundant moisture, droplets are repeatedly carried upward and downward by ascending and descending air currents. Hence, these drops quickly reach the required size for raindrops. Besides collision, electrical attraction between opposite charged particles also plays an important role in bringing about coalescence. This process is applicable only to those clouds which do not extend beyond the freezing level.

TYPES OF PRECIPITATION

All precipitation occurs from clouds. It is possible if atmosphere is unstable. The most important process of cloud formation is through adiabatic cooling of an upward moving air mass. Therefore, precipitation types are based on the conditions under which large masses of moist air are induced to rise to higher elevations. There are three primary modes for uplift of the air and each of these produces its own characteristic type of precipitation. Based on these modes for uplift of the air, precipitation types can be described as convective, cyclonic and orographic precipitation.

Convectional precipitation

Thermal convection of warm and moist air masses creates this type of precipitation. It occurs when the earth's surface within a conditionally unstable or moist atmosphere becomes heated more than its surroundings leading to significant upward motion. It gives maximum rainfall with minimum cloudiness. Convective rain or showery precipitation occurs from convective clouds, *i.e.*, cumulonimbus or cumulus. It falls as showers with rapidly changing intensity. Convective precipitation falls over a certain area for a relatively short time, as convective clouds have limited horizontal extent. Most precipitation in the tropics appears to be convective. Convective precipitation is more intense and of shorter duration.

Orographic precipitation

Orographic or relief precipitation is caused when masses of air pushed by wind are forced up the side of elevated land formations, such as large mountains. The lift of the air up the side of the mountain results in adiabatic cooling, and ultimately condensation and precipitation. In mountainous parts of the world subjected to relatively consistent winds (*i.e.*, the trade winds), a more moist climate usually prevails on the windward side of a mountain than on the leeward side. Moisture is removed by orographic lift, leaving drier air on the descending, leeward side where a rain shadow is observed. In India, southwest monsoon gives copious rainfall on the windward slope of mountains, whereas on the leeward side there are extensive rain shadow regions. The Western ghats, the Himalayas and the Aravallis are the best examples why rainfall is copious towards windward.

Cyclonic precipitation

Cyclonic or frontal precipitation is produced when two opposing air currents with different temperature meet followed by vertical lifting which gives rise to condensation and precipitation. When a low pressure cell is developed near the surface, winds in the lower atmosphere tends to converge to the centre of the low from all directions. Horizontal movement towards a common location implies an accumulation of mass called convergence. It gives rise to vertical motions near the centre of low to carry away the accumulated mass. A cyclonic storm is a large, low pressure system.

A boarder region between two adjacent air masses having different characteristics with respect to temperature and humidity is called a front. When masses of air with different density meet, warmer air overrides colder air, causing precipitation. In turn, precipitation can enhance the temperature and moisture contrast along a frontal boundary. Frontal systems generally surround extratropical cyclones or lows, which form when warm air meets cooler air. Frontal precipitation falls out of nimbostratus clouds. Warm fronts occur where the warm air scours out a previously lodged cold air mass. The warm air overrides the cooler air and moves upward. Warm fronts are followed by extended periods of light rain and drizzle. This is because, while being lifted the warm air gets cooled due to its expansion and forms clouds that leads to precipitation. Cold fronts occur when a mass of cooler air dislodges a mass of warm air. This type of transition is sharper, since cold air is denser than warm air. The rain duration is less, and generally more intense, than that which occurs ahead of warm fronts.

FACTORS AFFECTING DISTRIBUTION OF PRECIPITATION

Different areas of the earth's surface receive different amounts of precipitation. The major factors that affect the type, occurrence, and amount of precipitation include moisture content in the atmosphere, air temperature, latitude, altitude, barometric pressure, air mass movement, differential cooling and heating of the atmosphere, atmospheric disturbances, physical barriers, landforms, and wind speed and direction. However, all these factors are not independent of each other. Each factor is dependent on others in many ways. Since these climatic factors are found in more than one combination over different parts of the earth, the pattern of precipitation distribution is rather complex.

Latitudinal belts of atmospheric convergence and divergence has maximum influence over the distribution of precipitation. Convergence zones are the belts of maximum precipitation as there is an upward movement of the air with the adiabatic cooling. Minimum precipitation occurs near the subsidence or divergence zones. The equatorial belt which is the convergence zone of north and south trade winds, receive maximum precipitation. Along with the seasonal shifting of doldrums, the trade wind convergence zone moves north or south of equator. In subtropical high pressure belts subsidence of air gives rise to anticyclonic conditions on the surface of the earth. Due to atmospheric stability, it has minimum precipitation. In the belt of westerlies, the cold polar winds meet the moist tropical air mass along the polar fronts and give rise to appreciable annual average precipitation in this mid-latitude convergence zone. In the belt of polar easterlies, there is much less precipitation due to subsidence. The low temperature also diminishes the moisture holding capacity of the air. In the cold temperate and polar zones, precipitation occurs mainly as snow.

Distribution of land and water imposes an important influence upon precipitation in the middle and high latitudes. In the same latitudes, the continents receive less precipitation than the oceans, for continental air masses are drier than maritime air masses. Interior locations on continents, being far away from the source of oceanic moisture, receive relatively less precipitation than coastal areas. Warm ocean currents are conductive to greater precipitation on the adjacent lands, whereas the cold currents produce only fogs. The effect of continents on the distribution of precipitation is very distinct in the vast land mass of Asia. The southwest monsoon gives more rainfall in the areas adjacent to the ocean, *i.e.*, west and east coast of Indian subcontinent and much less inwards. The Deccan plateau remains much drier than the coasts.

Mountain barriers force the moisture bearing winds to ascend and cause heavy rains. All over the world, wherever the prevailing winds blowing from the oceans to the land are blocked by mountain ranges or plateaus, heavy rainfall results on the windward slopes while the leeward side remains dry.

The average annual precipitation over the whole earth is 113 cm, distributed unevenly. The equatorial regions and monsoon areas of south-east Asia record the highest rainfall. The temperate regions receive moderate amount of precipitation. The dry regions of subtropical high pressure belt and polar regions receive little precipitation.

Environmental factors that affect the composition of precipitation include air pollution. For instance, one of the leading causes of acid rain is nitric oxide and sulfur dioxide gases released in car exhaust and factory smoke.

RAINFALL CLIMATOLOGY

Study of rainfall over a long period is usually referred to as rainfall climatology. It gives information on general patterns and characteristics of rainfall of a particular region that helps in understanding rainfall characteristics in a better way so that efficient and high productive cropping systems may be planned for that region. It may also help in taking decisions on optimum time of agricultural operations like sowing, intercultural operations, fertiliser application, scheduling irrigation, plant protection, harvesting, post harvesting operations, *etc.* for different crops. Package of practices for a crop may be developed for the places or regions with similar rainfall characteristics (such as amount, intensity and distribution) by grouping them.

Total rainfall or its mean over a certain period is of little use in crop production. Mean values are frequently inflated by a few heavy and extreme events, or outlier events, that may have occurred, a phenomenon which is especially prevalent in dry regions or in dry months. Rainfall of a place can be meaningfully described by its intensity, duration and frequency. Rainfall intensity is the rate at which rain falls; duration of rainfall is the time for which rain is falling and frequency of rainfall is the number of times it falls. Some other important rainfall characteristics that are relevant to crop production are rainfall probability, variability and dependability.

The World Meteorological Organisation has divided a calendar year into 52 weeks called standard meteorological weeks of 7 days each. The first week starts on 1st January and so on, the last week being of 8 days. In a leap year, the 9th week has 8 days.

Total and mean total rainfall

The rainfall is recorded on daily basis. Thus its weekly, monthly, seasonal and annual total can be calculated for analysis. The amount of rainfall collected by a raingauge in 24 hours is known as daily rainfall and summing up of daily rainfalls for one year, the total amount is known as annual rainfall. The mean of the annual rainfalls over a period of 35 years is known as mean annual rainfall or normal annual rainfall. Annual rainfall in a particular year may be more or less than the annual mean value. The ratio of actual rainfall in a particular year at a given place to the normal annual rainfall at that place is called as index of wetness and expressed in percentage. Index of wetness gives an idea of wetness of a year, indicating deficiency of rainfall. A 70% index of wetness means a rainfall deficiency of 30%. A deficiency of 30 to 45% is known as large deficiency, 45 to 60% as serious deficiency and more than 60% as disastrous deficiency.

Coefficient of variation of annual precipitation

The average amount of precipitation need not necessarily be a constraint in successfully carrying out an agricultural operation. In fact, one can adapt his crop management practices if he has the assurance that the rains will fall when needed or as expected. Rainfall variability can be assessed through coefficient of variance (CV). The CV can be calculated by dividing standard deviation by mean rainfall and expressed in per cent. It indicates whether the region has a high or low rainfall in a particular year. It can also be used for relative comparisons of rainfall variability between one region and the next. The CV of annual precipitation is an index of climatic risk, indicating a likelihood of fluctuations in reservoir storage or crop yield from year to year. Considering the annual CV, the India Meteorological Department has defined to assess a particular year as normal, deficit or scarce. When CV is –19 to 19% of annual normal rainfall, it is a normal year with respect to receiving annual rainfall. When the value is –20 to –59% it is deficit year and when it is –60% or negatively more it is scarce.

Distribution of rainfall

Rainfall received during specified intervals like week, month or season indicate its distribution. This can be assessed by the rainy days, dry spells and wet spells. Distribution of rainfall is more important than total rainfall in a season for crop production. Yield is not directly proportional to the amount of rainfall. If the rainfall received in a particular day is equal to or more than 2.5 mm, it is called a meteorological rainy day and a day receiving more than 5 mm rainfall is called as crop rainy day. Number of rainy days indicates rainfall distribution in a season.

Effective rainfall

The amount of rainfall that is received during a cropping season may not be entirely used by the crop. Part of it may be lost through surface runoff, deep percolation beyond the root zone or by evaporation of rain intercepted by foliage. The extent of its utility in any given region determines rainfall efficiency. A fraction of the total rainfall is only available for consumptive use of crop. Thus, effective precipitation (ER) is the amount of precipitation that is actually added and stored in the soil for use of the crop. During drier periods, less than 5 mm of daily rainfall would not be considered effective, as this amount of precipitation would likely evaporate.

Rainfall characteristics like amount, intensity and distribution; land and soil characteristics like topography, soil texture, structure, bulk density, permeability and infiltration; groundwater contribution and crop characters like root zone depth, degree of ground cover, growth stage, *etc.* are the major factors that influence the effective rainfall. Crop management practices like tillage, bunding, ridging, terracing, mulching, *etc.* minimise the surface runoff, thus increasing effectiveness of rainfall. Frequent light rains intercepted by foliage with full ground cover are close to 100% effective. The rain water must penetrate to a depth of at least 10 to 12 cm with an estimated rainfall of 15 - 20 mm in a single occasion. The annual effective rainfall commonly amounts to about one-third of total rainfall.

PRECIPITATION AND CROP PRODUCTION

Precipitation, especially rain, has a dramatic effect on agriculture. All plants need at least some water to survive, therefore rain being the most effective means of watering is important to agriculture. Without precipitation, this planet would be an enormous desert. The amount and duration of precipitation events affect both water level and water quality in a water body on the land. A regular rain pattern is usually vital for the crops, however, too much or too little rainfall can be harmful, even devastating to crops. Plants need varying amounts of water to survive.

Importance of water on crop plants

1. Water has high solvent power and this plays an important role in crop plants as the plants get their nourishment from soil only in solution form.
2. Water plays an important role in life processes of crop plants.
3. The heat capacity of water is high and its high thermal stability helps in regulation of the temperature of crop plants.
4. Water has the highest heat conduction capacity and due to this the heat produced by the activity of a cell is conducted immediately by water and distributed evenly to all plant parts.
5. The viscosity of water is higher than that of many solvents and this property helps in protecting the crop plants and trees against mechanical disturbances.
6. The transparency of water facilitates the passage of light to great depths and this helps for the survival of aquatic plants.
7. The high surface tension of water helps in movement of water into and through the plant parts.
8. Rainfall influences the distribution of crop plants in particular and vegetation in general, as the nature of vegetation of a particular place depends on the amount of rainfall.

Influence of rainfall on crop production

Rainfall characteristics such as intensity, duration and frequency of rainfall, number of rainy days and effective precipitation have significant influence on crop production. Cropping system can be planned by understanding the rainfall pattern. Rainfall analysis helps in taking decisions on time of sowing, scheduling of irrigation, time of harvesting, *etc.* Rain analysis is necessary for designing farm ponds, tanks of irrigation projects. The success and productivity of crops, particularly dryland crops depend upon moisture availability, length of growing season, crop water requirement and cropping systems.

Soil moisture regime during crop growing period has greater importance in dryland crops in *kharif* season. Productivity of crops depends on the distribution and intensity of rainfall during growing period of the crops rather than the total precipitation received. Moisture retentive capacity of soil and evapotranspiration demand of the atmosphere and crops are the other two major parameters that affect the productivity of crops. Choice of crops and crop varieties, and cropping systems is guided by rainfall distribution and its intensity during different stages of crop growth and soil characteristics. If potential evapotranspiration (PET) is twice the evapotranspiration (ET) for a period of twelve weeks or more, rice can be grown under rainfed conditions. Where rainfall is more than 1200 mm, rice can be grown rainfed as the length of water availability period is about 13 weeks. With annual rainfall of 380 to 600 mm and humid period length of 2–4 weeks, only short duration pulse crops up to 12 weeks duration can be grown rainfed.

Length of growing season (LGS) is defined as the length of rainy season, plus the period for which soil moisture storage at the end of rainy season and post rainy season and winter precipitation can meet the crop water requirements. The LGS vary from 90 days in north-western India to 300 days in north-eastern region. It varies from 120 to 150 days in semiarid regions and from 150 to 180 days in sub-humid regions. Water balance approach estimates actual evapotranspiration (AET) on monthly or weekly basis. Index of moisture adequacy (IMA) is the ratio of AET and PET. Based on this index water availability period during crop growing season has been classified as dry (IMA 0 to 0.24), semidry (0.25 to 0.49), sub-moist dry (0.5 to 0.74), moist (0.75 to 0.99) and humid (IMA more than 1.0). Growing period commence when IMA is 0.5 or more and extends till it reaches 0.25. Under the same rainfall distribution, the LGS may vary from region to region depending upon the soil depth and its water retentive capacity. Choice of crops for dryland conditions should be based on the humid period during crop season. Short duration crops of 10 to 12 weeks duration like pearlmillet, minormillets, greengram, blackgram, cowpea, mothbean, *etc.* can be taken for the arid regions having rainfall around 300 mm.

Chapter 10

Monsoon

The word 'monsoon' is derived from the Arabic word *mawsim* meaning season. The Arabs during their travel towards India recognised that by June strong southwesterly winds commence over the Arabian Sea and bring a lot of rains to the subcontinent. The term was first used in English in British India and neighbouring countries to refer to the big seasonal winds blowing from the Bay of Bengal and Arabian Sea in the southwest bringing heavy rainfall to the area. In the Indian subcontinent, monsoon is one of the oldest weather observations, an economically important weather pattern over June through September every year, and the most anticipated weather event and unique weather phenomenon. Due to its effects on agriculture, flora and fauna, and other economic, social, and environmental effects, a monsoon is keenly followed and studied. It has significant impact on the overall well-being of subcontinent residents.

The term 'monsoon' is customarily used in India to refer to the rainy season during the northern hemisphere summer from June to September and is traditionally defined as a seasonal reversing wind accompanied by corresponding changes in precipitation, but is now used to describe seasonal changes in atmospheric circulation and precipitation associated with the asymmetric heating of land and sea. Now the term monsoon has been broadened to include almost all of the phenomena associated with the annual weather cycle within the tropical and subtropical land regions of the earth.

Intergovernmental Panel on Climate Change (IPCC) describes monsoon as a tropical and subtropical seasonal reversal in both the surface winds and associated precipitation, caused by differential heating between a continental-scale land mass and the adjacent ocean. India Meteorological Department defines it as the seasonal reversals of the wind direction along the shores of the Indian Ocean, especially in the Arabian Sea, that blow from the southwest during one half of the year and from the northeast during the other half.

The definitions include major wind systems that change direction seasonally. The cyclic change of wind and rain can be separated into the summer monsoon period characterised by warm and humid southwest winds, and the winter monsoon period characterised by dry and cool northeast winds. In fact, monsoon systems are a major feature of the general circulation of the atmosphere in subtropical latitudes of most regions of the world.

Monsoons are large scale sea breezes which occur when the temperature on land is significantly warmer or cooler than the temperature of the ocean. These temperature imbalances happen because oceans and land absorb heat in different ways. The air temperature remains relatively stable over oceans for two reasons; water has a relatively high heat capacity, and both conduction and convection will equilibrate a hot or cold surface with deeper water up to 50 m. In contrast, land surface has lower heat capacities and it can only transmit heat into the earth by conduction and not by convection.

During warmer months land temperatures rise more quickly. As the land surface becomes warmer, the air above it expands and an area of low pressure develops. Meanwhile, the ocean remains at a lower temperature than the land, and the air above it retains a higher pressure. This difference in pressure causes sea breezes to blow from the ocean to the land, bringing moist air inland. This moist air rises to a higher altitude over land and then it flows back towards the ocean; thus, completing the cycle. However, when the air rises, and while it is still over the land, the air cools. This decreases the air's ability to hold water, and causes precipitation over the land. This is why summer monsoons cause so much rain over land. In the colder months, the cycle is reversed. Then the land cools faster than the oceans and the air over the land has higher pressure than air over the ocean. This causes the air over the land to flow to the ocean. When humid air rises over the ocean, it cools, and this causes precipitation over the oceans. The cool air then flows towards the land to complete the cycle. Most summer monsoons have a dominant westerly component and a strong tendency to ascend and produce copious amounts of rain because of the condensation of water vapour in the rising air. Winter monsoons, by contrast, have a dominant easterly component and a strong tendency to diverge, subside, and cause drought.

MECHANISM OF CREATING MONSOON

The scientific explanation of the Asian monsoon was first proposed by Edmund Halley in 1686 through his thermal concept. He described this circulation as a gigantic land-sea breeze and the land-sea temperature contrast as its key factor. During winter the landmass of Asia cools more rapidly than the surrounding oceans. This results in the development of a strong high pressure centre over the continent. On the other hand, the pressure over adjacent oceans is relatively low. Consequently the pressure gradient is directed from land to sea. Thus cold, dry air flows from land to the ocean. In summer the pressure and temperature conditions just get reversed and as per the pressure gradient the hot and moisture laden wind flows from sea towards land. This moist unstable wind accompanied by atmospheric turbulence or landform barrier rises up and brings the rain.

Hermann Flohn, a German climatologist proposed a new explanation for the monsoon in 1951. During the northern hemisphere summer all the wind belts are displaced towards the north and in the northern winter they shift to south following the apparent movement of the sun in its zenith. As a result, in July the mean zonal axis of the equatorial winds in its zenith coincides with 10° N. In January it coincides with 5° S. The subtropical highs are also displaced a little. Under this condition the intermediate regions between these planetary zones are encroached upon by two contrasting annual wind regimes. The monsoon thus represents the succession of these two wind regimes. Flohn's explanation thus ascribes the monsoon to the thermal response of the tropical continental atmosphere to be the annual variation of solar radiation.

It is now believed that thermally induced surface low and high pressure centres alone cannot produce the monsoon circulation. Recent theories have laid greater emphasis on the influence of Tibet Plateau and the jet streams on the origin of monsoonal circulation over the Indian subcontinent and its adjoining areas. The seasonal migration of the inter tropical convergence zone (ITCZ) or equatorial trough is mainly responsible for the seasonal reversal of the wind. December-January-February (DJF) is the winter season for the northern hemisphere. At that time the solar radiation reaching the northern latitudes are highly oblique, *i.e.*, the temperature is quite low. From the month of March onwards the earth reverses its position and the northern hemisphere moves nearer to the sun, indicating the beginning of summer in the northern hemisphere. Consequently, the land and water masses there record an increase in their surface temperature. Since the land mass gets heated more rapidly than the water, a low pressure zone is created readily on the subcontinent. With a gradual rise in the sea surface temperature (SST), the sea surface also has a low pressure zone, but less intense than that of the continental. This thermal equator so formed facilitates the migration of the ITCZ northward of its DJF position. The highest SST in the global ocean during March-April-May is observed over the northern Indian Ocean. SST and ITCZ are highly coupled and a fluctuation in the SST has a large impact on the system due to the changes occurring in evaporation rate. The higher the SST, the larger is the convective current and that in turn facilitates the formation of clouds. In ITCZ where the trade winds converge, the moist air above the oceans is forced to rise up. As it moves up, it starts getting cooled adiabatically. After attaining a certain height when the air gets cooled to dew point, and the relative humidity reaches 100%, the condensation starts and clouds are formed.

The variation in winds is associated with the movement of the ITCZ. The initiation of the south west summer monsoon is marked broadly by reversal of northeasterly winds into southwesterly winds. This distribution is strongly influenced by presence of the East African mountains.

INDIAN MONSOON

Monsoon in India can be categorised based on their spatial spread over the sub-continent into two branches, *viz.*, Arabian Sea branch and Bay of Bengal branch. Alternatively, it can be categorised into two segments based on the direction of rain bearing winds, south-west monsoon (SW monsoon) and north-east monsoon (NE monsoon). Based on the time of the year that these winds bring rain to India, they can also be categorised into two rain periods, the summer monsoon (May to September) and the winter monsoon (October to November).

Southwest monsoon

The northern and the central Indian subcontinent heats up considerably during the hot summers from March to June. This causes a low pressure area over the area. To fill this void, the moisture-laden winds from the Indian Ocean rush into the subcontinent. These winds, rich in moisture, are drawn towards the Himalayas. The Himalayas act like a high wall, blocking the winds from passing into central Asia, and forcing them to rise. As the clouds rise, their temperature drops and precipitation occurs.

The southwest monsoon is generally expected to begin around the beginning of June and fade away by the end of September. Monsoon arrival is very often a gradual beginning with a short period of transition from extreme heat to a more humid atmosphere with light rain. The normal date of onset of monsoon can be worked out at a location based on normal rainfall computed for five days period, commonly known as a 'pentad'. A sudden increase in normal rainfall of a particular pentad over one or two preceding pentads in May, June or July depending upon location, is considered the monsoon pentad for defining the onset of monsoon. The midday of the particular pentad is taken as the normal date of onset of monsoon. India Meteorological Department (IMD) has produced maps indicating the normal dates of onset of monsoon for the country as a whole. IMD also declares the actual onset of monsoon every year for the country.

The monsoon typically breaks over Indian territory by around May 25, when it lashes the Andaman and Nicobar islands in the Bay of Bengal. However, June 1 is regarded as the date of onset of the monsoon in India, the average date on which the monsoon strikes near the Malabar Coast of Kerala.

The standard deviation of onset date is 8 days which means that if the monsoon onsets between May 24 and June 9, it is considered normal. Onset before May 24 is early onset whereas onset after June 9 is called delayed or late onset.

The moisture-laden winds on reaching the southernmost point of the Indian peninsula, due to its topography, become divided into two parts, the Arabian Sea branch and the Bay of Bengal branch. The Bay of Bengal branch covers north eastern states while the southern peninsula is covered by the Arabian Sea branch.

The Arabian Sea branch extends towards a low-pressure area over the Thar Desert and is roughly three times stronger than the Bay of Bengal branch. It first hits the Western Ghats of the coastal state of Kerala around June 1. This branch of the monsoon moves northwards along the Western Ghats with precipitation on coastal areas, west of the Western Ghats. The eastern areas of the Western Ghats do not receive much rain from this monsoon as the wind does not cross the Western Ghats. By June 10, it reaches Mumbai and appears over Delhi by June 29. The Bay of Bengal branch initially tracks the Coromandel Coast, from Kanyakumari in Tamil Nadu to Odisha. It flows over the Bay of Bengal heading towards north-east India and West Bengal, picking up more moisture from the Bay of Bengal. Both the branches merge somewhere in north India by the end of June and travel together to northwest India by mid-July covering the whole country.

The winds arrive at the eastern Himalayas with large amounts of rain. Mawsynram, situated on the southern slopes of the Khasi Hills in Meghalaya, is one of the wettest places on earth. After the arrival at the eastern Himalayas, the winds turn towards the west, travelling over the Indo-Gangetic Plains, and pouring rain all along its way. The monsoon reaches extreme northwest (Rajasthan) by 15th July, taking 45 days to cover the whole India. The monsoon begins to withdraw early from Rajasthan by September 1. As India further cools during September, the southwest monsoon weakens. The withdrawal of monsoon is a far more gradual process than its onset. It withdraws from Mumbai by October 5. The monsoon usually withdraws from the whole country except the southern peninsula, by October 15. By the end of November, it leaves the country. The duration of monsoon over western Rajasthan is hardly 60 days while it is 165 days over Kerala.

Northeast monsoon

During the post-monsoon months of October to December, a different monsoon cycle, the northeast 'retreating' monsoon brings in dry, cool, and dense Central Asian air masses to large parts of India. Around September, with the sun fast retreating south, the northern land mass of the Indian subcontinent begins to cool off rapidly. With this, air pressure begins to build over northern India, the

Indian Ocean and its surrounding atmosphere still holds its heat. This causes cold wind to sweep down from the Himalayas and Indo-Gangetic Plains towards the Indian Ocean, south of the Deccan peninsula. They cross India diagonally from northeast to southwest. While travelling towards the Indian Ocean, the dry cold wind picks up some moisture from the Bay of Bengal and pours it over peninsular India and parts of Sri Lanka. Tamil Nadu and some parts of Kerala experience significant precipitation in the post-monsoon and winter seasons. About 50–60% of the rain received by the state of Tamil Nadu is from the northeast monsoon. Parts of West Bengal, Odisha, Andhra Pradesh, Telengana, Karnataka and north eastern states of India also receive minor precipitation from the northeast monsoon. The onset of northeast monsoon normally over south peninsula is on October 15 and the withdrawal is complete by December 1.

FACTORS CONTROLLING THE INDIAN MONSOON

Predicting the summer monsoon has been an important task for the India Meteorological Department (IMD). Apart from the north-south oscillation of the ITCZ, a number of synoptic scale disturbances, orographic features, ocean-atmosphere interactions as well as global climatic conditions contribute to the proper functioning of the monsoon. Snow and even the soil moisture content are also intricately related with the monsoon.

Ocean-atmosphere coupling

To maintain an approximate steady state climate the ocean and atmosphere move excess heat from the tropics to the heat deficit poles. The wind exerts a stress on the sea surface that induces wind driven circulations. Because of this interaction the influence of the annual cycle of the sea surface temperature (SST) is important in the seasonal movement of the convective zone. The seasonal mean SST patterns show that the warm waters are co-located with strong precipitation areas over the ocean. Warm SST occurs in the northern hemisphere mostly during June-July-August (JJA) season and mostly during December-January-February (DJF) in the southern hemisphere. During JJA, SST is greater than 28° C over the western Pacific (where ITCZ is located), Bay of Bengal and central Indian Ocean. The warm SST zone moves south in the following month with the warmest water. The organised convection is inhibited below the threshold temperature value of 28° C. Also if the SST exceeds the threshold, the convection is found to be associated more with the convergence of moist air rather than evaporation. In the land atmosphere interface heating gradients are associated with the monsoon precipitation. The monsoon rainfall in India is significantly reduced with cold SST over the Arabian Sea.

Orographic features

There are two major orographic barriers that appear to have significant influence on the atmospheric flow. These are the Himalaya-Tibetan Plateau complex and the Eastern Mountains of Africa. The Ghats and the Burma mountain ranges have also been found to influence the monsoon on local scale. The Himalayan mountain range acts as a giant geographical barrier to the cold continental winds of central Asia in its north during the northern winter and to the moisture laden southwesterlies during the northern summer. The Himalayan mountain is the key factor working for the maintenance of the Asian summer monsoon and also allow high temperatures on the Tibetan Plateau to extend low pressure towards the southwest. As a whole, this contributes to the strengthening of the Indian monsoon producing high rainfall over the subcontinent through mechanical/dynamical and thermodynamical mechanism.

The heavy rainfall in the Western Ghats on the other hand was also due to the topographically forced ascent. The downstream side of these ranges (*i.e.*, the leeward side) are devoid of rainfall owing to the high precipitation (and hence loss of moisture content) upstream. This describes the occurrence of drier climate of the peninsula.

Atmospheric systems over Bay of Bengal

The onset of monsoon is characterised by the presence of some synoptic scale disturbances. After onset of monsoon, the monsoonal rain gets highly intensified due to these disturbances. Cyclonic storms or disturbances are a regular and frequent feature of the whole southwest monsoon period. Atmospheric whirls of all degrees of intensity and magnitude pass up from the Bay of Bengal to the coastal districts and interior of India.

Depressions, the principal rain bearing system, generally form over the Bay of Bengal within the monsoonal trough. On an average 6–7 depressions form each year during JJAS. Year to year variation in their number however, is quite large. The energy supplied by the release of latent heat of condensation becomes the major energy source for these depressions. The depressions formed in early June are responsible for the advance of the southwest monsoon. In the months of July and August they usually form north of 18° N in the northwest Bay. By September its origin point shifts to the central Bay. The disturbances originating over the warm ocean water move onto the land, bringing heavy to exceptionally heavy rainfall. Generally an active spell of monsoon is characterised by high frequency genesis of the synoptic scale systems over the Bay, which migrates onto the subcontinent and produces the awaited monsoonal downpour. The zone of maximum rainfall shifts to northwest and then to northeast as the depression moves from northwest to north to northeast. After the break of monsoon, once again, these disturbances pacing westward, reintroduce the rainfall. The convection over the Bay is a vital factor for the formation and variation of the synoptic scale disturbances. On the other hand these disturbances apart from providing rainfall to the subcontinent do play a major role in transport of heat and moisture, thus maintaining the global meteorological balance.

El-Nino - Southern Oscillation effect

El-Nino is a 'warm' ocean current originating along the coast of Peru that replaces the usual 'cold' Peru or Humboldt Current. This warm surface water reaching towards the coast of Peru with *El-Nino* are pushed westwards by the trade winds. Thus, the temperature of the southern Pacific Ocean is raised. A reverse condition is known as *La Nina.*

Southern Oscillation, a phenomenon first observed by Sir Gilbert Thomas Walker, Director General of Observatories in India in 1920s, refers to the see-saw relationship of atmospheric pressures between Tahiti (an island in the central southern Pacific Ocean) and Darwin (the capital city of the Northern Territory, Australia situated on the Timor Sea, arm of the Indian Ocean). He noticed that when it was high pressure in Tahiti, it was low pressure in Darwin and *vice versa.* A Southern Oscillation Index (SOI), based on the pressure difference between Tahiti and Darwin, has been formulated by the Bureau of Meteorology (Australia) to measure the strength of the Oscillation. Walker noticed that the quantity of rainfall in the Indian subcontinent was often negligible in the years of high pressure at Darwin (and low pressure at Tahiti). Conversely, low pressure at Darwin predicts the precipitation quantity in India. Thus, he established the relationship of Southern Oscillation with quantities of monsoon rains in India.

Ultimately, it was realized that the Southern Oscillation is just the corresponding atmospheric component of the *El Nino/La Nina* effect (which happens in the ocean). Therefore, in the context of the monsoon, the two cumulatively came to be known as the ENSO. The ENSO is known to have a pronounced effect on the strength of southwest monsoon over India with the monsoon being weak (causing droughts in India) during the *El Nino* years whereas *La Nina* years had particularly good monsoon strength over India.

The correlative patterns between Indian west coast summer rainfall and *El-Nino/La-Nina* may have been stronger decades ago because the correlation coefficient between all-India summer monsoon rainfall and *El- Nino/La-Nina* shows 95% significance during the past 150 year except for the years after 1980.

Indian Ocean dipole effect

Although ENSO was statistically effective in explaining several past droughts in India, in the recent decades the ENSO-monsoon relationship seemed to weaken in the Indian subcontinent. The 1997 strong ENSO failed to cause drought in India. However, it was later discovered that just like ENSO was an event in the Pacific Ocean, a similar seesaw ocean-atmosphere system in the Indian Ocean was also at play. It was discovered in 1999 and named as the Indian Ocean Dipole (IOD). An index to calculate it was also formulated. IOD develops in the equatorial region of Indian Ocean from April to May peaking in October. With a positive IOD, winds over the Indian Ocean blow from east to west resulting in a much warmer Arabian Sea (western Indian Ocean near African coast) with a colder and drier eastern Indian Ocean around Indonesia. In the negative dipole year, the reverse happens making Indonesia much warmer and rainier. It was demonstrated that a positive IOD index often negated the effect of ENSO, resulting in increased monsoon rains in several ENSO years like the 1983, 1994 and 1997. Further, it was shown that the two poles of the IOD, the eastern pole (around Indonesia) and the western pole (off the African coast) were independently and cumulatively affecting the quantity of rains for the monsoon in the Indian subcontinent.

Snow

Snow (mainly the Eurasian, Tibetan and the Himalayan snow) is one of the surface boundary conditions that might have some influence on the monsoon predictability. The snow accumulation during the Himalayan winter and spring is inversely related to the subsequent summer monsoon rainfall over India.

Land surface

The monsoon may be associated with the fluctuations of surface temperature and also to other surface conditions, mainly the soil moisture.

CROPPING SEASONS IN INDIA

In agriculture, the growing or cropping season is the period of each year when crops can be grown. The agricultural crop year in India is from July to June. The Indian cropping season is classified into two main seasons; *kharif* and *rabi* based on the monsoon. The *kharif* cropping season is from July to October during the southwest monsoon and the *rabi* cropping season is from October to March (winter). The crops grown in between these two major seasons are called '*zaid*' crops.

Kharif crops are sown at the beginning of the southwest monsoon and harvested at the end of the southwest monsoon; sowing season is from May to July and harvesting season is September to October. The important *kharif* crops include rice, maize, sorghum, pearl millet, finger millet, arhar, green gram, blackgram, soybean, groundnut, sesame, cotton, jute, sunhemp, *etc.* *Rabi* crops need relatively cool climate during the period of growth but warm climate during the germination of their seed and maturation. Sowing time is October-December and harvesting time is from February to April. The *rabi* crops include wheat, barley, oats, chickpea, lentil, rapeseed and mustard, sunflower, linseed, potato, tomato, cabbage, pea, *etc.* *Zaid kharif* crops are sown in August-September and harvested in December-January. Important crops are rice, jowar, rapeseed, cotton, horsegram, *etc.* *Zaid rabi* crops are sown in February-March and harvested in April-May. Important crops include watermelon, cucumber, leafy and other vegetables.

Some parts of India have their own classification of cropping seasons. It is usually determined by climate and crop selection. However the monsoon plays a major role in the nomenclature of these cropping seasons, *i.e.*, pre-monsoon (March-May), monsoon (June-September), post-monsoon (October-November) and winter (December-February).

Pakistan and Bangladesh are two other countries that are using the term '*kharif*' and '*rabi*' to describe their cropping patterns. The terms '*kharif*' and '*rabi*' originate from Arabic language where *kharif* means autumn and *rabi* means spring.

IMPACT OF MONSOON ON INDIAN AGRICULTURE

Indian economy is considered as an agricultural economy. Indian agriculture is heavily dependent on the rains, for growing crops. The monsoon accounts for about 80% of the rainfall in India. Since over half of the Indian farmlands are rainfed, monsoon is critical to the food sufficiency for the country.

Therefore, the agricultural calendar of India is governed by monsoon. Good monsoon always means a good harvest and brings in cheers all around the country. Good monsoons correlate with a booming economy. Any fluctuations in the time distribution, spatial distribution or quantity of the monsoon rains may lead to conditions of floods or droughts causing the agricultural sector to adversely suffer. This has a cascading effect on the secondary economic sectors, the overall economy, food inflation and therefore the overall quality and cost of living for the general population in India.

Though, the improved irrigation and availability of electricity has reduced the dependence of agriculture on monsoon to a small extent, the hydroelectric power generation, ground water availability and its recharge are totally dependent on good monsoon rains.

The Asian drought of 1987 was one of the worst of this kind from Afghanistan to the Philippines. The physical cause of the drought was the weakness of the southwest monsoon circulation, the driving mechanism of the region's weather patterns. A below-normal rainfall and high temperature damaged crops and livestock throughout Asia. In India, *kharif* season grain and oilseed production reduced drastically below expected levels and winter crops that depend on residual summer moisture for germination were planted well beyond the normal time. While the implications of the drought did not approach the famine years of the 1970s, commodity imports rose, reserves and exportable supplies were reduced, and growth was limited in various economic sectors as the agricultural sector accounts for nearly 30% of India's gross domestic product and employs almost 67% of the labour force.

Chapter 11

Weather Hazards

A weather hazard is an extreme weather event that threatens people or property. There are various weather hazards that affect agricultural productions. These hazards vary from one climatic zone to another and occur at different times, although some regions are more vulnerable to certain hazards than other. These weather hazards depend on the latitude of the place on the world map. These hazards affect crops and agricultural productivity negatively if they are not checked. Every agricultural zone has one particular hazard that is associated with it and in most cases affects the crop productions and reduces the annual crop yield. Weather hazards include cyclones, droughts, floods, hailstorms, heat-waves and cold-waves.

The hazard is distinguished from an extreme event and a disaster. An extreme event is simply an unusual event; it does not necessarily cause harm. A disaster is an event that does cause harm in significant amounts. Weather hazards are often called as natural hazards. The natural hazards become natural disasters when people's lives and livelihoods are destroyed.

Hydrologic hazards are the result of either excess or a severe lack of water, *i.e.*, flood or drought. Hydrometeorological hazard is a process or phenomenon of atmospheric, hydrological or oceanographic nature that may cause loss of life, injury or other health impacts, property damage, loss of livelihoods and services, social and economic disruption, or environmental damage. Human and material losses caused by natural disasters are a major obstacle to sustainable development. By issuing accurate forecasts and warnings in a form that is readily understood and by educating people about how to prepare against such hazards, before they become disasters, lives and property can be protected.

DROUGHT

Drought is a relative term used with reference to deficiency of rainfall when compared to normal rainfall of a given location. It is an extended period when a region experiences a deficiency in its water supply, whether atmospheric, surface or ground water. However, drought cannot be defined in terms of amount of rainfall or number of days without rain. Since ancient times droughts have had far-reaching effects on humankind by causing the failure of crops, decreasing natural vegetation, and depleting water supplies. Drought is different from other hazards, in that it develops slowly. It is a hazard that requires many months to emerge and that may persist for many months or years thereafter. This type of hazard is known as a 'creeping' hazard. Generally, this occurs when a region receives consistently below average precipitation.

Classification of drought

Drought results from a long continued dry weather and/or insufficient rain, which causes loss of soil moisture, depletion of underground water supply and reduction of stream flow (NDMA, 2010). Drought is frequently defined according to disciplinary perspective, where droughts are related to precipitation (meteorological), streamflow (hydrological), soil moisture (agricultural) or any combination of the three (Dracup *et al*., 1980). It can also be classified based on duration and nature of users. However, in both the types of classification, demarcation between any two types is not well defined and at times there is overlapping of the cause and the effect of one on the other.

The National Commission on Agriculture in India defines three types of droughts; meteorological, hydrological and agricultural depending upon the degree of deficiency of rainfall or soil moisture. A drought is said to be meteorological if the rainfall deficiency is more than 25% of the normal rainfall over an area. When deficiency is between 25 to 50%, it is called moderate drought and the term severe drought is used for rainfall deficiency of more than 50%. However, most locations around the world have their own definition of meteorological drought based on the climate normals in the area. The country as a whole may have a normal monsoon, but different meteorological districts or subdivisions can have below normal rainfall. The rainfall categories for smaller areas are defined by their deviation from normal rainfall of a meteorological area. Meteorological drought usually precedes the other kinds of drought.

A prolonged meteorological drought can result in hydrological drought with marked depletion of surface water and consequent drying up of reservoirs, lakes, rivers and fall in the ground water table. Hydrological drought tends to show up more slowly because it involves stored water that is used but not replenished. Hydrological drought may occur even when the rainfall is normal, if there has been a substantial reduction in surface water holding capacity.

Agricultural drought occurs when soil moisture and rainfall are inadequate during the growing season to support healthy crop growth. This condition can also arise independently from any change in precipitation levels when soil conditions and erosion triggered by poorly planned agricultural endeavours cause a shortfall in water available to the crops. There are three different types of agricultural drought; permanent, seasonal and contingent. Permanent drought occurs in arid areas where there is not enough precipitation to satisfy the water needs of plants. Such droughts are noticed in western Rajasthan and Gujarat and agriculture is impossible without irrigation throughout the crop season. In most parts of tropics, seasonal drought is noticed with well defined wet and dry seasons. Drought can be expected every year owing to seasonal variations in atmospheric circulation patterns. Contingent drought is the characteristic of the humid and sub-humid areas and occurs over a period of time when the rains fail. It is the serious hazard to agriculture in India because of the unpredictability of the monsoon rain. Depending on the arrival and behaviour of the monsoon during the crop season, contingent drought can further be divided into early, mid and late season droughts. In India, crops are selected across the country to fit the normal rainfall conditions. If there is any deviation in normal rainfall towards deficit, it adversely affects the crop's growth, development and production under rainfed conditions. The delay in monsoon (early season drought) adversely affects sowing area of different crops across the country reducing the cropped area. Mid and late season droughts also affect the food grain production. The late season drought is known as terminal drought.

Agricultural drought can also be categorised into three types; atmospheric, soil and physiological droughts. Low humidity with hot and dry desiccating winds cause atmospheric drought even under conditions of adequate moisture in the soil. Wilting of crops occur due to an underdeveloped root system or the physical inability of roots to draw water to compensate the water loss from the leaves during the stress period. But wilting due to atmospheric drought is always reversible. Soil drought occurs when soil moisture content lags behind evapotranspiration at times of greatest need of water by plants, such as during grand growth period. Physiological drought occurs when plant root systems fail to absorb water from the soil due to increased concentration of soluble salts in soil solution. Presence of soluble salts in soil water in high concentration reduces the availability of water to the plants, although sufficient water is in the soil. Presence of certain chemicals in soil solution may also restrict absorption of water by the root system.

The India Meteorological Department (IMD) defined drought on the basis of rainfall received and the area affected due to drought during the monsoon season. If the rainfall over the country is less by 10% of the long period average and 20% of the cropped area is affected due to drought, it is known as an all-India drought. As per IMD, there are 31 such recorded drought years since 1875; 1877, 1891, 1899, 1901, 1904, 1906, 1907, 1911, 1913, 1916, 1918, 1920, 1926, 1930, 1941, 1951, 1965, 1966, 1968, 1972, 1974, 1979, 1982, 1986, 1987, 2002, 2012, 2014, 2015 and 2017. The worst drought year was 1918 when nearly 70% of the country was affected by drought.

Causes of drought

Drought can be caused by a number of factors. The most important one though relates to the amount of water vapour in the atmosphere as it creates precipitation. Generally, rainfall depends on the amount of water vapour carried by regional atmosphere, combined with the upward forcing of the air mass containing that water vapour. If these combined factors do not support precipitation volumes sufficient to reach the surface, the result is a drought. More rain occurs where there are moist, low pressure air systems. If there are dry, high pressure air systems, less moisture is available to produce precipitation. This results in a deficit of water for the areas over which they move.

Human activities like over farming, excessive irrigation, deforestation, and soil erosion can also directly trigger droughts. Widespread cutting down of trees for fuel or for clearing forests for agriculture reduce the soil's ability to hold water that causes desertification leading to drought. Constructing a dam on a large river may help provide electricity and water to irrigate farmland near the reservoir. However, it may also cause drought downstream by severely reducing the flow of water.

Common causes of drought in India include inadequate monsoon rainfall, high temperature coupled with high wind speed causing higher evapotranspiration, inadequate water availability due to high water loss in storage, distribution and utilities, shift in agricultural practices (from low to moderate and to high water demanding crops), high rate of increase in human and animal population in a specific location, and over exploitation of surface and ground water.

Impacts of drought

A drought is caused for an extended period of months or years when a region notes a deficiency in its water supply. Although droughts can persist for several years, even a short, intense drought can cause significant damage and harm the local economy. One of the sectors where the immediate impact of drought is felt is agriculture. The impact of water deficit is also felt in ground water recharge, non-availability of quality seeds, land degradation and reduced power supply.

The impact of a drought may be either direct or indirect and varies in nature as well as intensity. The extent and intensity of drought impact is determined by prevailing economic conditions, the structure of the agricultural sector, management of water resources, food grain reserves, *etc.* The direct impact of drought is generally classified under four categories, *viz.* physical, economic, social and environmental. The relative and absolute magnitudes of each impact will however, depend on specific regional characteristics. Droughts cause a loss of assets in form of crops, livestock, fisheries and productive capital as these are immediate consequences of water shortage (NDMA, 2010). Many of these losses are then passed on to consumers in the form of higher commodity pricing. The social impacts include increased chance of conflict over commodities, fertile land, and water resources. Other social impacts include abandonment of cultural traditions, loss of homelands, changes in lifestyle, and increased chance of health risks due to poverty and hygiene issues. The environmental impacts of drought include loss in species biodiversity, migration changes, reduced air quality, and increased soil erosion.

Influence of drought on crop plants

Water stress due to drought has a tremendous adverse effect on crop plants. The physiological and morphological characters of a plant are influenced by the water balance in the plant system. The crop plants vary in critical growth periods with respect to water requirement, and water stress during these critical stages would drastically reduce the biomass production. During flowering stage, a little stress of moisture by virtue of drought substantially reduces the size of inflorescence thereby affecting the final yield. Water shortage at fertilisation and grain filling also influence the final yield of any crop.

Water constitutes about more than 80% of the fresh weight of the herbaceous plants and 50% of the woody plants. Cell division and enlargement are very sensitive to drought stress. Water content in the protoplasm below a certain level results in stunted growth or even kills the plant. Drought also affects nutrient absorption, carbohydrate and protein metabolism and translocation of ions and metabolites. Protein breakdown injures the drought stressed plant due to the accumulation of toxic products of protein breakdown such as ammonia. Abscission of leaves, fruits and seeds are induced by plant water deficit during droughts.

Drought preparedness and mitigation

Drought preparedness, coupled with appropriate mitigation actions and programmes, can reduce and, in some cases, eliminate many of the impacts associated with drought. Drought planning process has three components; monitoring and early warning, risk assessment, and mitigation. The monitoring and early warning component of a drought plan essentially provides the foundation on which timely decision can be made by decision makers at all levels, from farmers to national policy makers. Given drought's creeping characteristics, monitoring the hydrological system can detect drought's early onset and its potential impacts. There should be interdisciplinary approach to reduce and mitigate the impacts of drought. The meteorological and hydrological observations should be systematically collected and processed to support and strengthen the scientific networks for the enhancement of technical capacities in meteorology, hydrology and other related fields.

Conservation practices minimise the disruption of the soil's structure, composition and natural biodiversity, thereby reducing erosion and soil degradation, surface runoff, and water pollution. The established practices of soil and water conservation are crop rotation, contoured row cropping, terracing, minimum tillage practices, erosion control structures, water retention and detention structures, windbreaks and shelterbelts, litter management, and reclamation of salt-affected soil. Soil and water conservation can be approached through agronomic and engineering measures. Agronomic measures include contour farming, off-season tillage, deep tillage, mulching and providing vegetative barriers on the contour. These measures prevent soil erosion and increase soil moisture. Alternate land use system like agroforestry, *i.e.*, integrating tree crops with annual herbaceous crops, may be encouraged. Carefully planned crop rotation can help to minimise erosion and allow farmers to plant less water-dependent crops in drier years.

Engineering measures differ with location, slope of the land, soil type, and amount and intensity of rainfall. These include soil and water conservation through regulated runoff, collection of surplus rain water, checking evaporational and seepage loss of water and recycling of collected water for life saving irrigation. These can be achieved by making water harvesting structures, scooping the land, opening ridges and furrows, contouring across the slope, and compartmental bunding, *etc.* Contour bunds, trenches and stone walls can prevent soil erosion and obstruct the flow of runoff. The retained water increases soil moisture and recharges the groundwater. Check dams (check dams are temporary structures constructed with locally available materials) and other gully-plugging structures can be constructed to restrict the runoff and soil erosion. Percolation ponds can be constructed by excavating a depression to form a small reservoir, or by constructing an embankment in a natural ravine or gully to form an impoundment. These features store water for livestock and recharge the groundwater. Rainwater can be collected from roofs or other suitable catchments and stored for later use. The wastewater (sewage) can also be recycled by proper treatment and purification. Water-supply projects can also be implemented for drought mitigation, with a view to strengthen drought preparedness. Activities such as water-use planning, rain-water harvesting, runoff collection using surface and underground structures, improved management of channels and wells, exploration of additional water resources through drilling and dam construction, are implemented as a part of a drought-mitigation plan.

The physiological approaches include application of suitable chemicals in crops as antitranspirants, growth retardants, and radiation reflectants for reducing the transpiration rate, checking the shoot growth and encouraging the root growth, and reducing the radiation, respectively which ultimately may reduce the impact of drought on crops. Genetic approaches for reducing the adverse impacts of drought aimed at developing new plant types, known as 'ideo types' which possess wider adaptability, shorter growth duration, drought resistance and high genetic yield potentiality. Plant breeders have developed many short duration varieties for various crops. These varieties are characterised by effective and extensive deep root system, drought tolerance, resistance to diseases, effective photosynthetic behaviour, dwarf plant types with lesser number of erect leaves and high yield potentiality. Thus, the drought can be evaded or escaped by substituting the traditional long duration varieties with these ideo types.

FLOODS

A flood is an overflow of water that submerges land which is usually dry. Flooding may occur as an overflow of water from water bodies, such as a river or lake, in which the water overtops, resulting in some of that water escaping its usual boundaries. Floods are recurrent phenomena since time immemorial. Almost every year some parts of the world or the other are affected by the floods of varying magnitude. Even in the same country, while some parts are suffering from devastating floods, other parts experience severe drought. Flood is the single most frequent disaster faced not only in India but by various parts of the world.

Causes of flood

The years in which actual rainfall exceeds the normal value by twice the standard deviation or more are defined as the years of flood. Most often floods are a secondary hazard resulting from other meteorological processes, such as prolonged rainfall, localized and intense thunderstorms, or onshore winds.

Inadequate capacity of the rivers to contain within their banks the high flows brought down from the upper catchment areas following heavy rainfall, leads to flooding. In coastal areas, floods are caused by cyclones and typhoons. Other causes include backing up of waters in tributaries at their outfalls into the main river often with synchronisation of floods in them, and landslides blocking stream courses resulting in the backwater overflowing river banks. However, the root cause of flood is excessive rainfall which occurs mainly in the monsoon months. Because of the varying rainfall distribution, many a times, areas which are not traditionally prone to floods also experience severe inundation. Areas with poor drainage facilities get flooded by accumulation of water from heavy rainfall. Flood may also occur due to an accumulation of rainwater on saturated grounds. Excess irrigation water applied to command areas and increase in ground water levels due to seepage from canals and irrigated fields also are factors that accentuate the problem of water logging. The problem is exacerbated by factors such as silting of the riverbeds, reduction in the carrying capacity of river channels, erosion of beds and banks leading to changes in river courses, obstructions to flow due to landslides, synchronisation of floods in the main and tributary rivers and retardation due to tidal effects (NDMA, 2008).

Flash floods

Flash floods occur with little or no warning, and are the result of intense rainstorms within a brief period of time. These cause high damages because of their suddenness. Flash floods occur in hilly and sloping lands where heavy rainfall and thunderstorms or cloudbursts are common. The phenomenon of sudden unprecedented heavy rain due to some peculiar climatic conditions is known as cloudbursts. Depression and cyclonic storms in the coastal areas of Odisha, West Bengal, Andhra Pradesh, Karnataka, and Tamil Nadu also cause flash floods. The states of Arunachal Pradesh, Assam, Odisha, Himachal Pradesh, Uttarakhand, the Western Ghats in Maharashtra and Kerala are more vulnerable to flash floods caused by cloudbursts. Sudden release of waters from upstream reservoirs, breaches in landslide dams and embankments on the banks of the rivers leads to disastrous floods.

Effects of floods

The primary effects of floods include loss of life, damage of crops and livestock, loss of properties, damage to buildings and other structures including bridges, roads and transport infrastructure, sewerage systems, roadways, canals, and disruption of power supply. Flood waters typically inundate farm land, making the land unworkable and preventing crops from being planted or harvested, which can lead to shortage of food both, for humans and farm animals. Entire harvests for a region may be lost in extreme flood circumstances. Some tree species may not survive prolonged flooding of their root systems. Floods are followed by epidemics. Supply of drinking water and restoration of sanitation are post flood challenges. Secondary effects of flooding include coastal erosion and soil erosion. Economic hardship due to a temporary decline in tourism, rebuilding costs, or food shortage leading to price increases is a common after-effect of severe flooding. The flood problem in India is mostly confined to the states located in the Indo-Gangetic plains, northeast India and occasionally in the rivers of central India.

Frequent or smaller floods can also bring many benefits, such as recharging groundwater, making soil more fertile and increasing nutrients in some soils. Flood waters provide much needed water resources in arid and semiarid regions where precipitation are very unevenly distributed throughout the year. Flooding can spread nutrients to lakes and water reservoirs, which can lead to increased biomass and improved fisheries for a few years.

Flood preparedness and mitigation

Floods being the most common natural disaster, people have, out of experience, devised many ways of coping with them. However, encroachments into the floodplains over the years have aggravated the flood problem and a need to take effective and sustained flood management measures has been felt (NDMA, 2008).

Flood forecasting has been recognised the most important reliable and cost effective non-structural measure for flood mitigation. Forecasting of floods allows for precautions to be taken and people to be warned so that they can be prepared in advance for flooding conditions. A long time-series of historical data that relates stream flows to the measured past rainfall events can have fairly accurate flood forecasts. Coupling this historical information with real-time knowledge about volumetric capacity in catchment areas, such as spare capacity in reservoirs, groundwater levels, and the degree of saturation of area aquifers is also needed in order to make the most accurate flood forecasts. Radar estimates of rainfall and general weather forecasting techniques are also important components of good flood forecasting. The intensity and height of a flood can be predicted with plenty of lead time, if good quality data is available. In India, the Central Flood Board was set up in 1954. The India Meteorological Department (IMD) and Central Water Commission (CWC) are working in close co-ordination for issue of flood forecasts. IMD has 10 flood meteorological offices and CWC has 20 flood forecasting divisions located in different flood prone river basins in the country.

Various mitigation measures can be taken to reduce the adverse impact of floods. In order to achieve control over the flow and height of the water carried by the river, the channel, floodplain or watershed must undergo some physical alterations.

1. Dams, retention basins or reservoirs on mainstreams or tributaries can be constructed to store excessive water and release it gradually after the threat is over.
2. Levees or floodwalls can be constructed to confine flood waters to a floodway, thereby reducing flood damage.

3. Channel improvement works can be taken up to increase the efficiency of drainage system. These include straightening to remove undesirable bends, deepening and widening to increase size of waterways; clearing to remove trees and other obstructions; lining with concrete to increase efficiency.
4. Watershed treatment can also be employed to help the soil on slopes to become more absorbent of rainfall until flood height is receded. Watershed treatment involves crop rotation, construction of terrace, contour strip cropping; selective planting and reforestation.
5. Any type of construction or land filling should not be allowed on the lateral boundaries of the designated floodways. This is done to ensure that the flow of water is not obstructed.

Other mitigation measures include change in crop management practices, flood insurance, temporary evacuation of persons, livestock and property from the path of flood waters, *etc.* The crops and varieties which can withstand water logging conditions for a fairly long period should be preferred in the flood prone areas.

TROPICAL CYCLONE

Major storms in the tropics are known as tropical cyclone. Depending on its location and strength, a tropical cyclone is referred to by names such as hurricane, typhoon, tropical storm, cyclonic storm, tropical depression, and simply cyclone. All forms of storm fall under the definition of tropical cyclone which is a synoptic scale cyclonic circulation with organised convection and surface wind circulation originating over tropical waters. In meteorology, a cyclone is an area of closed, circular fluid motion rotating in the same direction as the earth. This is usually characterised by inward spiralling winds that rotate anti-clockwise in the northern hemisphere and clockwise in the southern hemisphere of the earth. Tropical cyclones are characterised by large low pressure centres and numerous thunderstorms that produce strong winds and heavy rain. Tropical cyclones feed on heat released when moist air rises, resulting in condensation of water vapour contained in the moist air. While tropical cyclones can produce extremely powerful winds and torrential rain, they are also able to produce high waves and damaging storm surge as well as spawning tornadoes. They develop over large bodies of warm water, and lose their strength if they move over land. This is why coastal regions can receive significant damage from a tropical cyclone, while inland regions are relatively safe from receiving strong winds. Heavy rains, however, can produce significant inland flooding, and storm surges can produce extensive coastal flooding up to 40 kilometres from the coastline.

Tropical cyclones are classified based on intensity into three main groups; tropical depressions, tropical storms, and a third group of more intense storms, whose name depends on the region. For example, a tropical violent windstorm in the north-western Pacific referred to as a 'typhoon'; in the northeast Pacific or in the Atlantic, it is called a 'hurricane' and in the southern hemisphere or the Indian Ocean basins, storms of tropical nature are referred simply as 'cyclones'. A tropical depression is an organised system of clouds and thunderstorms with a defined, closed surface circulation and maximum sustained winds of less than 63 km/hour. A tropical storm is an organised system of strong thunderstorms with a defined surface circulation and maximum sustained winds between 63 and 117 km/hour. A hurricane or typhoon or tropical cyclone is a system of sustained winds with a speed of at least 119 km/hour. Storm surge is the abnormal rise of sea level which occurs when the cyclone moves from ocean to continent. It is a disastrous and destructive feature of cyclone, which could cause massive destruction in the coastal region and misery to the coastal human population and livestock. The intensity of tropical cyclones is expressed by using Saffir-Simpson scale as given in table 11.1.

Table 11.1 Saffir-Simpson scale to express intensity of tropical cyclones

Category	Intensity	Pressure (mb)	Wind speed (km/hour)	Storm surge (m)	Damage
1	Minimal	> 980	119–154	1.0–1.7	Damage to shrubs and trees; no major damage to most building structures; flooding of low-lying coastal areas
2	Moderate	965–979	155–178	1.8–2.6	Considerable damage to shrubs and some trees being blown down; damage to door and windows of residences; flooding of coastal roads and low-lying areas
3	Extensive	945–964	179–210	2.7–3.8	Large trees blown down; some structural damage to small buildings; flooding of sea-level coastland 10 km or more inland
4	Extreme	920–944	211–250	3.9–5.6	Many shrubs, trees are blown out; severe damage to roofing structures, windows, and doors of buildings; flooding of low-lying areas as much as 13 km inland; major damage to structures near shore due to battering by waves

Category	Intensity	Pressure (mb)	Wind speed (km/hour)	Storm surge (m)	Damage
5	Cata-strophic	< 920	> 250	> 5.6	All shrubs, trees blown out; complete failure of roofs on residences and industrial buildings; overturning/ sweeping away of small buildings; evacuation of all residential areas on low-lying ground within 16 km of shore

(*Source*: https://www.nws.noaa.gov)

Cyclones and their impact in India

Indian subcontinent is the worst affected region of the world, having a coast line of 7516 km; 5400 km along the mainland, 132 km in Lakshadweep and 1900 km in Andaman and Nicobar Islands. The states Andhra Pradesh, Odisha, Tamil Nadu, West Bengal and the union territory Puducherry on the east coast and Gujarat on the west coast are more vulnerable to cyclone disasters. Cyclones occur in the month of May-June and October-November, with primary peak in November and secondary peak in May. Although cyclones affect the entire coast of India the east coast is more prone compared to the west coast.

Cyclones are known to cause severe damage to infrastructure through high speed winds. Very strong winds accompanying a cyclonic storm damage trees, installations, residences, communication systems, *etc*., which results in loss of life and property. Torrential rainfall (more than 30 cm/hour) associated with cyclones is another major cause of damages. Unabated rain gives rise to unprecedented floods. Heavy rainfall from a cyclone is usually spread over wide area and cause large scale soil erosion and weakening of embankments. A severe tropical cyclone causes storm surge; as a result of which sea water inundates low lying areas of coastal regions drowning human beings and livestock. This also erodes beaches and embankments, destroys vegetation and leads to reduction of soil fertility.

Table 11.2 Damage potential of cyclonic disturbances

System (Intensity in km/hour)	Damage expected	Action suggested
Deep depression (52-61)	Minor damage to loose and unsecured structures	Fishermen advised not to venture into the seas
Cyclonic storm (62-87)	Damage to thatched huts; breaking of tree branches causing minor damage to power and communication lines	Total suspension of fishing operations

System (Intensity in km/hour)	Damage expected	Action suggested
Severe cyclonic storm (88-117)	Extensive damage to thatched roofs and huts; minor damage to power and communication lines due to uprooting of large avenue trees; flooding of escape routes	Total suspension of fishing operations; coastal hutment dwellers to be moved to safer places; people in affected areas to remain indoors
Very severe cyclonic storm (118-167)	Extensive damage to *kutcha* houses; partial disruption of power and communication line; minor disruption of rail and road traffic; potential threat from flying debris; flooding of escape routes	Total suspension of fishing operations; mobilise evacuation from coastal areas; judicious regulation of rail and road traffic; people in affected areas to remain indoors
Very severe cyclonic storm (168- 221)	Extensive damage to *kutcha* houses; some damage to old buildings; large-scale disruption of power and communication lines; disruption of rail and road traffic due to extensive flooding; potential threat from flying debris	Total suspension of fishing operations; extensive evacuation from coastal areas; diversion or suspension of rail and road traffic; people in affected areas to remain indoors
Super cyclone (222 or more)	Extensive structural damage to residential and industrial buildings; total disruption of communication and power supply; extensive damage to bridges causing large scale disruption of rail and road traffic; large-scale flooding and inundation of sea water; air full of flying debris	Total suspension of fishing operations; large-scale evacuation of coastal population; total suspension of rail and road traffic in vulnerable areas; people in affected areas to remain indoors

(*Source*: Attri and Tyagi, 2010)

Recurring frequency of cyclones leads to drastic transformation in the pattern of coastal land use by turning fertile areas into saline wastelands due to marine incursions. In addition, standing crops are destroyed; salt deposits from saline water render farms unproductive and even barren. An entire cropping season is lost and areas affected by severe storm surge are sometimes not cultivable during subsequent seasons also. Very often harvested food crop stocks are also destroyed. Horticultural crops suffer heavy losses. Livestock including farm and dairy cattle, poultry, sheep and goats also suffer.

Although tropical cyclones are known for destruction they cause, they may be important factors in the precipitation regimes of places they affect, as they may bring much-needed precipitation to dry regions, relieving drought conditions. Tropical cyclones also help maintain the global heat balance by carrying warm, moist air from the tropics to the temperate latitudes and polar regions. This helps to maintain equilibrium in earth's troposphere and maintains relatively stable and warm temperature worldwide.

Cyclone preparedness and mitigation

Although it is not possible to completely avoid natural disasters, their effects can be minimised by taking some known long and short term structural and non-structural mitigation measures such as developing proper early warning systems, creating awareness at all levels in the concerned communities, coastal afforestation, construction of shelters, embankments, dykes, coastal roads, bridges, canals, *etc.* through better preparedness, mitigation measures and improved response mechanisms (NDMA, 2008).

The India Meteorological Department (IMD) has introduced a 4-stage warning system as described below to issue cyclone warnings to the disaster managers.

Pre-cyclone watch: The pre-cyclone watch is issued when a depression forms over the Bay of Bengal irrespective of its distance from the coast and is likely to affect Indian coast in future. It is issued at least 72 hours in advance of the commencement of adverse weather. It is issued at least once a day.

Cyclone alert: The cyclone alert is issued at least 48 hours before the commencement of the bad weather when the cyclone is located beyond 500 km from the coast. It is issued every three hours.

Cyclone warning: The cyclone warning is issued at least 24 hours before the commencement of the bad weather when the cyclone is located within 500 km from the coast. Information about time and place of landfall are indicated in the bulletin.

Post landfall outlook: The post landfall outlook is issued 12 hours before the cyclone landfall, when the cyclone is located within 200 km from the coast. More accurate and specific information about time and place of landfall and associated bad weather is indicated in the bulletin.

The sector that is most affected by tropical cyclones is agriculture and food security. The sharpest falls in agricultural production has been recorded during those years that have experienced the occurrence of severe cyclones in India. Various options have been conceptualised for protecting the crops (NDMA, 2008). The crop calendar for major crops should be adjusted to normal cyclone periods. The cultivation of high yielding varieties and hybrids of rice in the coastal areas, which can withstand surge waters, floods and cyclone winds without damaging the grain and straw quality, should be encouraged. Short height crops should be grown to avoid lodging in cyclone prone areas. The crops can be protected from floods caused by cyclonic rainfalls through improvement of drainage systems by strengthening of bunds all along canals and tanks; de-silting of water reservoirs, canals and drains; widening and deepening of drains and canals; keeping reservoirs and canals weed free; and diversion of existing drains and canals and alignment of new canals and drains keeping in view that water will be flowing out freely during disaster periods. The drains which are directly joining the sea should be maintained regularly.

The animals and birds should be shifted to a higher and safer place as soon as warnings for impeding cyclones are received. Cowsheds should be constructed in a way that they can withstand strong cyclone winds. Sufficient nutritious feed should be preserved for the bad weather periods. The animals and birds should be vaccinated beforehand against communicable diseases during emergency periods. The farmers should be advised to insure their crops and livestock. Contingent planning should be made sufficiently ahead. The farmers may be provided with the required inputs like seeds/ seedlings and fertilisers timely to go for a contingent crop to compensate the crop loss due to cyclonic storms and storm surges.

HAILSTORMS

Any thunderstorm which produces hail that reaches the ground is known as a hailstorm. Hail is a form of solid precipitation. It consists of balls or irregular lumps of ice, each of which is called a hailstone. Hailstones consist mostly of water ice and measure between 5 millimetres and 15 centimetres in diameter. Hail is composed of transparent ice or alternating layers of transparent and translucent ice at least 1 millimetre thick, which are deposited upon the hailstone as it travels through the cloud, suspended aloft by air with strong upward motion until its weight overcomes the updraft and falls to the ground. Stones larger than 2 cm are usually considered large enough to cause damage.

Hail is possible within most thunderstorms as it is produced by cumulonimbus clouds. It is most common within continental interiors of the mid-latitudes, as hail formation is considerably more likely when the freezing level is below the altitude of 3400 m. Movement of dry air into strong thunderstorms over continents can increase the frequency of hail by promoting evaporational cooling which lowers the freezing level of thunderstorm clouds giving hail a larger volume to grow in. Hail is less common in the tropics despite a much higher frequency of thunderstorms than in the mid-latitudes because the atmosphere over the tropics tends to be warmer over a much greater altitude. In the tropics, hail formation tends to be confined mainly to higher elevations.

Agriculture is typically the most affected by hailstorms because it causes severe crop damage and even a minor storm with relatively small size hailstones can have a devastating effect on agriculture. Hailstorms cause fruit injury, canopy damage besides uprooting and breakage of plants in the horticultural crops like pomegranate, grapes, citrus, mango, banana, papaya, onion and vegetables. Though in some cases the losses are irreparable, in others where the damage is less the crop can be saved. The extent of losses varies from mild to high depending upon the force and size of hail that falls, and intensity, duration and time of occurrence of hailstorm during the growing season.

Various preventive measures can be taken to minimise the damage by hailstones. The formation of hailstones or their size can be reduced through cloud seeding. This may be done by injecting sufficient ice nuclei like silver iodide into the cloud environment so that the hailstones are prevented from being large enough to hit the crops with high force, or even they get melted before reaching the ground. However, the success of cloud seeding to prevent hailstorms depends upon the accuracy in forecasting of hailstorms with sufficient lead time for placement of seeding materials in the right quantity at the right place. Weather satellites and weather radar imagery are very useful tools to detect the presence of hail-producing thunderstorms. The radar data has to be complemented by a knowledge of current atmospheric conditions which can allow one to determine if the current atmosphere is conducive to hail development.

Fixing hail damage on plants is not always possible. The best approach is to clean up the debris and trim off broken stems and leaves. The hail damage to trees may require pruning away of the most affected branches. Injuries to minor trees will heal. But fungicide should be applied immediately to prevent rot in the injured branches. Open wounds on the growing shoots and branches present entry points for bacterial and fungal pathogens and also insects. Some plants may be too heavily affected and fixing hail damage is not possible. These plants should be removed and replaced. Short duration fruit crops like banana and papaya which suffer from more than 50% permanent damage to the plants should be replanted. Mango, pomegranate, grapes and citrus fruit crops do not have permanent loss. The remedial measures for these crops depend upon the vegetative and reproductive growth stage of crops and the extent of damage. Physical barriers such as hail nets or other protective screens can be used in the orchards of these high value fruit crops as a preventive measure.

HEAT-WAVES

A heat-wave is a prolonged period of excessively hot weather, which may be accompanied by high humidity. A heat-wave is generally measured relative to the usual weather in the area and relative to normal temperature for the season. Temperature that is considered to be normal by people from a hotter climate area can be termed a heat-wave in a cooler area if they are outside the normal climate pattern for that area. The definition of heat-wave recommended by the World Meteorological Organisation is applicable when the daily maximum temperature of more than five consecutive days exceeds the average maximum temperature by 5 °C, the normal period being 1961-1990.

As per the India Meteorological Department, heat-waves are not to be considered as serious till maximum temperature of a station reaches at least 40° C for plains and 30° C for hilly regions. In a station with normal maximum temperature of less than or equal to 40° C, if the prevailing maximum temperature is 5 to 6° above the normal, then heat-wave is presumed to exist there. If the departure from normal is 7° or more, then the term severe heat-wave is used. Where the normal maximum temperature of a station is more than 40° C, heat-wave is declared when departure from normal is by 4 to 5° and severe heat-wave is declared when departure from normal is by 6° or more. When actual maximum temperature remains at 45° C or more irrespective of normal maximum temperature, heat-wave is declared. Heat-waves develop in May and June in northern and central India. Rajasthan and Punjab experience heat-waves due to their being in the interior. The strong north-westerly winds along the land trajectory over hot regions inhibit the setting in of the sea breeze over the eastern coastal belt. Thus, there is high incidence of heat-waves and their persistence over coastal Odisha and Andhra Pradcsh. Chennai becomes another centre of heat-waves in July and August.

Excessive heat affects the development of field crops like rice, wheat, maize, potato and soybean, and also many fruit crops like apples, grapes, *etc.* Yield of many crops may be reduced by high heat when it occurs at a critical time in the plant's development. Blowing of hot winds leads to pollen injury and premature fruit drop. High temperatures, especially during both day and night, occurring during the pollination of maize can significantly reduce yields. Soybeans exposed to heat stress during flowering can result in pollen sterility and reduced seed set. Heat stress during the ripening stage of wheat can affect grain quality and yield. Shade structures may be made to protect vegetable crops grown on sandy soils during summer. Windbreaks can also reduce the effect of heat-wave by decreasing the velocity of the hot winds.

Livestock heat stress is a serious concern for livestock producers. Livestock such as poultry are severely impacted by heat-waves. Milk production and cattle reproduction also decrease during heat-waves. Livestock producers need to be proactive before and during a heat-wave to prevent the effects of heat stress. Electric fans and coolers can be used in the shelters during summer to save the animals from heat-wave.

COLD-WAVES

Cold-waves are defined with respect to the normal minimum temperature experienced over a given area. According to India Meteorological Department, if the normal minimum temperature of a station is equal to or more than 10° C, then cold-waves and severe cold-waves are defined when there is a marked fall of temperature by 5 to 6° and by 7° or more, respectively. At the station with normal minimum temperature less than 10° C, cold-waves and severe cold-waves are defined when there is a marked fall of temperature by 4 to 5° and by 6° or more, respectively. Nearly 75% of cold-waves in the country occur during January to February. Excluding the extreme south peninsula, the country is susceptible to cold-waves. The north western part of the country and areas further north are more favourable regions for the genesis of cold-waves and the incidence of 5 to 20 days of frost in each month during the cold season is common. The damage by cold-waves is more in the low lying areas, where cold-waves settle and remain for a long time on the ground.

Ground frost which adversely affects various crops is experienced under cold-waves. Cold-waves that bring unexpected freezes and frosts during the growing season in mid-latitude zones can kill plants during the early and most vulnerable stages of growth, resulting in crop failure as plants are killed before they can be harvested economically. Night irrigations can be applied to the crops like potato and tomato to mitigate the effect of frost and severe cold. Smokes and heaters are used in the fruit orchards to reduce the effect of cold and frost. Straw mats, screens of dry grass netting, plastic covers can be placed over small plants during late afternoon and removed early in the morning for protection against frost. A cold-wave can cause death and injury to livestock and wildlife. Electric heaters can be used to elevate the shelter temperature during strong cold.

AGRICULTURE AND WEATHER RELATIONS

Agriculture is an economic activity that is highly dependent upon weather and climate in order to produce the food and feed necessary to sustain human life. Despite technological improvements that increase crop yields, extreme weather events or weather hazards have caused significant yield reductions in some years at different regions. For any ecosystem, the interrelation between biotic factors (living things like plants, microorganisms, animals, *etc.*) and abiotic factors (temperature, sunlight, rainfall, wind) is necessary. Similar is the case for agriculture. The interrelationship between agriculture and abiotic factors is unavoidable. It may lead to both positive and negative impact in crop production.

The cropping pattern, selection of crops, cultural operations, application of farm inputs, harvesting and even storage and transport is decided by the prevailing weather conditions. Some of the examples are:

- Farmers start ploughing their fields only after soil has enough moisture.
- Operations like application of pesticides or herbicides are withheld if rains are expected.
- Farmers harvest at early morning to avoid moisture loss in produce due to bright sunshine.
- Harvesting of matured cereals, pulses or oilseeds is delayed if rains are forecasted or already harvested crops are taken to roofed threshing floor before rains cause any damage.

There are many examples that establish relationship between weather and agriculture.

- Sunshine, wind movement, temperature has direct relation with photosynthesis, transpiration and respiration activity.
- Flowering is influenced by duration of sunlight and darkness, termed as photoperiodism.
- Some plants need precipitation for fruit setting and for flowering. A pre-monsoon shower sometimes encourages early ripening of mangoes. In south India, it is called as 'mango showers'.
- No urea is topdressed before expected heavy rainfall and no topdressing after storm either. In the former case it saves urea from being lost in the environment and that later helps in managing bacterial diseases.

Chapter 12

Climatic Normals for Crops and Livestock Production

Different crops are grown over a wide range of agroecosystems. Though the crops are raised in varied climates, ranging from very hot to very cold, and very wet to very dry, across temperate, tropical and semiarid zones, they are very sensitive to climate. The crop production is dependent on climatic parameters like rainfall, light, temperature, relative humidity and carbon dioxide concentration. Each kind of cultivated crop plants has its own optimum growth requirements. Breeding and selection of new cultivars have allowed for a greater adaptability to less favourable growing conditions than was possible in the past, but the inherent climatic requirements of a specific kind of crop have not changed materially.

CLIMATIC NORMALS FOR CROPS

Crop management is very much about managing climate risk so as to have financially viable and sustainable agricultural systems. Professor Henry Nix of the Australian National University posed three questions that need to be addressed when considering what crop can be grown where (Nix, 1981).

1. For any given crop which areas offer the greatest biophysical advantages?
2. For any given area which crops offer the greatest advantages?
3. For any given crop and area how much productivity be raised and sustained?

The first considerations for answering to these questions are the climatic determinants which may influence the distribution of cropping and other land use patterns. Weather induced variability in crop productivity accounts for about two-thirds while other factors including soil and nutrient management account for one-third of the productivity. Thus a sound knowledge on climatic requirements of the intended crops is required for successful crop production. If the crops and cultivars are not well adapted to the area where they are grown, their cultivation in that area may be uneconomical.

Rice (*Oryza sativa*)

Rice crop is best suited to tropical and sub-tropical humid climate but it is grown in varieties of climates except extreme cold temperate conditions. The climatic factors that affect rice production are temperature, day length and humidity. Rice crop requires about 1200 to 1800 mm water. Therefore, if this much rain occurs during crop season with uniform distribution will be sufficient for crop. An average monthly rainfall of 200 mm is required to grow the low land rice and 100 mm to grow upland rice successfully. Soil moisture stress from flowering to milk stage leads to more than 75% reduction in grain yield. Moist humid weather during vegetative growth and dry-sunny weather during ripening is most desirable. A relative humidity of 60–80% is said to be optimum. Flowering is inhibited by relative humidity below 40%.

The atmospheric temperature has considerable effect on growth and development of rice plant. Rice needs relatively high temperature for their optimum growth and development. Temperature requirement of rice varies according to different growth stages. The crop can be grown successfully where average air temperature is more than 21° C for 4 to 5 months. Low temperatures inhibit tillering and prolong the crop duration. For vegetative growth a temperature range of 25 to 30° C and for flower initiation 27 to 29° C is ideal. A temperature range of 20 to 25° C is the best for grain filling and ripening. Temperature below 14° C and above 38° C may induce sterility. Night temperature of 23° C is ideal during ripening phase. High temperature especially during night leads to loss of reserved food through greater respiration. Temperature of water in the rice field should not be less than 21° C or more than 31° C at any time during the crop season.

Rice crop prefers brighter and prolonged sunshine for enhanced photosynthetic activity and higher yield. It should receive more than 300 sunshine hours during the last 45 days before harvest. Low sunshine during vegetative stage may not significantly affect the grain yield. However, low solar radiation would affect grain filling and it may increase chaffy grain production. High wind velocity or storm causes lodging or shattering depending upon the crop growth stage. Gentle wind is the best for rice cultivation because the supply of carbon dioxide and its utilisation is regulated to the maximum.

Table 12.1 Critical temperature of rice crop for different growth stages (Yoshida, 1978)

Growth Stages	Critical temperature (in degree Celsius)		
	Low	High	Optimum
Germination	16–19	45	18–40
Seedling emergence and establishment	12–35	35	25–30

Growth Stages	Critical temperature (in degree Celsius)		
	Low	High	Optimum
Rooting	16	35	25–28
Leaf elongation	7–12	45	31
Tillering	9–16	33	25–31
Panicle initiation	15	—	
Panicle differentiation	15–20	30	—
Anthesis	22	35–36	30–33
Ripening	12–18	> 30	20–29

Wheat (*Triticum aestivum*)

Wheat crop can be grown in tropical, subtropical and temperate zones. The cool weather for vegetative phase and warm weather for maturity is very conducive to a good crop of wheat. Wheat plants are sensitive to very cold or frost injury at any stage of growth especially at reproductive stage, if temperature is below 15° C. The ideal temperature requirement varies with plant type and stage of growth. A mean maximum temperature of 25° C and mean minimum of 12° C is considered optimum for grain development. The critical mean temperature requirements of the plant at different stages are given in the table below.

Table 12.2 Critical mean temperature requirement of wheat crop

Growth stages	Mean daily temperature (°C)
Germination	20–25
Tillering	16–20
Accelerated growth	20–23
Grain filling	23–25

Table 12.3 Adverse effect of temperature on wheat crop

Temperature (°C)	Effect on wheat crop
20 to 22	most optimum
> 22	irregular germination with frequent death of embryo
35	endosperm decomposes by microbial activities
16 to 22	active vegetative growth
22 to 42(during growth)	height, root length and tiller count is reduced
> 25 (during grain formation)	shrivelling of grains causing reduction in test weight

Well distributed rainfall of 350 to 400 mm during the crop period or four irrigations, one at crown root initiation stage and the remaining three at 35 to 40 days interval produce optimum yield. Wheat crop is a long day favouring crop. However, the new cultivars are less photosensitive.

Maize (*Zea mays*)

Maize is a warm weather crop. It is grown under extremely divergent climatic conditions ranging from tropical to temperate region. The most suitable temperature for germination is 21° C and for growth is 32° C. It can be successfully grown where the night temperature does not go below 15° C as crop stops growing below this level. The crop cannot withstand frost at any stage of its growth. Young plants are highly susceptible to hailstorm.

A well distributed rainfall of 500 to 750 mm during the cropping season is required for a successful crop. Moisture requirement of the crop is low from germination to initiation of reproductive stage. Maize is resistant to drought and can recover from early season drought. It requires huge water at reproductive stage. Soil moisture stress prior to silking, during silking and after silking reduces grain yield by 25, 21 and 50%, respectively (Reddy and Reddy, 2007). On the other hand, as water logging is hazardous for the crop at any growth stage. The maize crop needs bright sunny days for an accelerated photosynthetic activity and rapid growth of plants but this has to be associated with an abundant water supply failing which the crop will start wilting or withering. Prolonged cloudy period is harmful for the crop but an intermittent sunlight and cloud or rain is the most ideal for its growth. Rain free sunny days during flowering period favour better pollen dispersal and seed set.

Sorghum (*Sorghum bicolour*)

Sorghum requires warm climate but can be grown under a wide range of conditions. The crop can tolerate high temperature better than any other cereal crop throughout its lifecycle. Minimum temperature for germination is 7 to 10° C, while optimum is 18 to 21° C. Minimum temperature for growth is 15° C, while the optimum is 27 to 30° C. Night temperature greater than 21° C favours cell elongation and delays floral initiation. Sorghum cannot recover from freezing temperatures except at juvenile stage. Flowering is hastened by short days and delayed by long days. The time of heading is influenced by temperature and photoperiod.

Sorghum can tolerate drought conditions very well because it remains dormant during moisture stress conditions but resumes growth when favourable conditions reappear. It can also tolerate water logging conditions better than any other cereal crop except rice. A minimum of 300 to 400 mm rainfall is required during cropping season. However, sorghum can successfully be grown in areas having average annual rainfall between 600 and 1000 mm.

Pearl millet (*Pennisetum glaucum*)

Pearl millet can germinate over a wide range of temperature from 15–40° C. A temperature of 30–35° C is optimum for growth and development.

The maximum rate of photosynthesis occurs between 35 and 40° C. It is a warm weather crop suitable for areas with 400 to 750 mm annual rainfall. The crop is resistant to drought conditions. During the vegetative growth period moist weather is useful but the rainfall at flowering time is harmful as it washes off the pollen and consequently there is poor seed setting.

Finger millet (*Eleusine coracana*)

Finger millet can be grown in tropical and subtropical climates. It requires 350 to 400 mm rainfall during the cropping season and can be cultivated in areas having annual rainfall up to 1000 mm. In regions of higher rainfall it can be raised on well drained soils as a transplanted crop.

Barley (*Hordeum vulgare*)

Barley is a crop of temperate climate, but it can be grown under subtropical climatic condition. It thrives best in areas having cool dry winters with low rainfall. The crop can withstand cool humid and warm dry climate but not warm humid climate. It cannot tolerate frost at any stage of growth and incidence of hailstorm at the grain development stage damages the crop completely. Rain during the growth period results in good growth of crop, but rain at maturity causes discolouration of the grains, thus rendering them unfit for malting. Intermittent drought during the growth period results in premature ripening and grains with high nitrogen content are unfit for malting. Uniform moisture supply and bright sunshine at the ripening stage are important for the production of bright kernels required by the malting industries. The barley crop requires around 12–15° C temperature during growth period and around 30° C at maturity. High temperature affects barley at various stages of growth. Higher temperature is known to induce abnormalities.

Bengal gram (*Cicer arietinum*)

Chickpea or bengal gram is a subtropical crop. It requires cool dry climate, frost-free with dew. Though it is essentially a winter season crop, severe cold and frost are injurious to it. The crop is very sensitive to excess moisture, high humidity and cloudy weather, which limit flower production, seed set and yield. It cannot tolerate heavy rains, hence is unsuited for humid tropics. Chickpea is grown on residual soil moisture where the rainfall ranges from 400 to 700 mm. Excessive rains soon after sowing or at flowering and fruiting or hailstorms at ripening cause heavy loss. It cannot withstand water logging even for a short period. The optimum temperature for germination is 15 to 20° C. The average air temperature varies from 25 to 30° C for good crop growth. Nights must be warmer with 20 to 25° C temperatures.

Pigeon pea (*Cajanus cajan*)

Pigeon pea or arhar grows well in warm tropical and subtropical climate. The crop prefers a fairly moist and warm climate during the period of its vegetative growth. The optimum temperature for germination is 19 to 43° C. The seeds do not germinate if the temperature is below 7° C and beyond 45° C. Growth increases with increase in temperature up to 35° C. The crop requires bright sunny weather during the flowering and ripening stages. It is highly susceptible to frost at the time of flowering. Cloudy weather and excessive rainfall at flowering damage the crop to a great extent. Most pigeon pea growing regions in India receive 600 to 1400 mm annual rainfall.

Lentil (*Lens culinaris*)

Lentil is adapted to cool growing conditions, and the young plants are tolerant of spring frosts. The optimum temperature for growth and development of the crop is 15 to 20° C. Low temperature less than 10° C delays germination with reduced vegetative growth. Lentil can be grown on stored soil moisture with annual rainfall of 350 to 500 mm rainfall. High humidity and excessive rainfall during the crop season encourages vegetative growth, which prevents good yield and can reduce seed quality. Excessive drought and high temperatures during the flowering and pod-fill period also reduce yields.

Blackgram (*Phaseolus mungo*)

As tropical crop blackgram tolerates high temperature. It needs a hot and humid growing season. It is a short-day plant, but day neutral cultivars are available for cultivation in the long-days of summer. The optimum temperature for better growth ranges between 25 and 35° C and it can tolerate up to 42° C. It is a hardy and drought resistant plant, and can be grown in areas receiving moderate to low rainfall. Heavy rains during flowering stage are harmful and adversely affect the production. It is sensitive to cloudy weather and cannot tolerate frost.

Greengram (*Phaseolus aureus*)

Greengram is a tropical pulse crop largely grown under semiarid and subtropical environment. It is well suited to the areas having an annual rainfall of 600 to 1000 mm. It requires hot climate and can tolerate drought to a great extent. Greengram can be grown in the temperature regime of 20 to 40° C, the optimum being 28 to 32° C. The crop is sensitive to low temperature and below 10° C germination fails.

Pea (*Pisum sativum*)

Pea is grown in varied climatic conditions. It requires cold and dry climate. The longer cold spell helps in increasing yield. Pea seed can germinate even at a minimum temperature of 5° C but the process is slow. The optimum temperature for germination is about 22° C. The optimum mean temperature for good growth is between 10 and 18° C. High temperatures are injurious to the crop. Frost can damage the plants during flowering stage. Pea is most sensitive to moisture stress at flowering stage.

Soybean (*Glycine max*)

Soybean is a tropical crop but can be grown in subtropical and temperate regions. It is relatively resistant to low and very high temperature. It grows well at 15 to 32° C temperatures. The crop does not grow if the temperature falls below 10° C but at the same time if it goes above 40° C the crop gets a greater setback in its growth, flowering, and seed formation or even in the seed quality. Flowering is affected below 20° C. The crop requires about 600 to 650 mm annual rainfall. High rainfall is harmful because higher moisture at germination causes rotting of seeds and seedlings and at later stages of growth, excess water causes anaerobic conditions, poor nodulation and poor growth and yield of the crop.

Groundnut (*Arachis hypogaea*)

Groundnut is essentially a tropical plant. It requires a long and warm growing season. Temperature is a major environmental factor that determines the rate of crop development. Minimum temperature for germination is 10° C, while the optimum is 20° C. The optimum mean air temperature for vegetative growth is 25 to 30° C. Temperature below 15° C and above 35° C inhibits the growth of groundnut. For reproductive growth the optimum temperature range is 20 to 26° C. Flowering may go satisfactorily up to a temperature of 33° C, but pod formation retards above 26° C.

A well distributed rainfall of at least 500 mm during growing season is required for a successful crop. The crop can be grown successfully with a rainfall up to 1250 mm. Soil moisture stress at various growth stages brings variation in crop duration. Early moisture stress followed by adequate water supply may increase crop duration and pod yield but stress at flowering, pegging and pod formation stages reduces the duration and pod yield is affected adversely. The groundnut crop cannot withstand frost for long, severe drought and water stagnation.

Rapeseed and mustard (*Brassica* spp.)

Rapeseed and mustard crops are grown in both tropical and temperate zones and require cool and dry weather for satisfactory growth. In India these crops are cultivated during winter season. The crop needs 18 to 25° C temperature for better growth. These crops are not drought tolerant. The crops grow well with 350 to 450 mm rainfall. However, high rainfall, high humidity and cloudy weather are not good for the crops. The crops cannot withstand water logging. About 10 bright sunshine hours per day are required for growth and development. Excessive cold and frost are also harmful to the crops.

Sesame (*Sesamum indicum*)

Sesame is grown in tropical and subtropical areas. It is basically a crop of the warmer climate, though its cultivation extends also to cooler climate as available in the Himalayas up to 1500 m altitude. A temperature of 25 to 27° C is ideal for germination, initial growth and flower formation. Germination is delayed below 20° C temperatures. The plants grow luxuriantly in temperature range of 24 to 32° C. Low temperatures at flowering results in pollen sterility and temperatures above 40° C at flowering seriously affect fertilisation. Sesame is susceptible to water logging, heavy continuous rainfall, frost and hailstorm at any stage of growth. But it is a fairly drought resistant crop.

Sunflower (*Helianthus annuus*)

Sunflower crop requires a cool climate during germination and seedling growth followed by warm weather up to flowering and maturity. Mean daily temperatures for good growth are between 18 and 25° C. Ideal temperature for fruit setting is 21 to 24° C. The crop requires bright sunshine for higher yields. Sunflower is a photoperiod insensitive crop. A good crop can be harvested with well distributed rainfall of 500 to 750 mm. Rainfall more than 750 mm, particularly on black soil with poor drainage affect the seed yield adversely. High humidity accompanied with cloudy weather at the time of flowering results in poor seed set.

Safflower (*Carthamus tinctorius*)

Safflower is considered to be a cool season crop, though it is adapted to wide range of climatic conditions. It is more or less day neutral but thermosensitive. The optimum temperature for seed germination is about 15.5° C. Seedlings are tolerant to low temperature. High temperature at the time of flowering is harmful to the crop. Day temperature in the range of 24–32° C at flowering is the best. When the flower buds are being formed temperature below zero degree Celsius may cause considerable damage in the form of sterile heads. Safflower is a drought resistant crop and grows well in areas receiving 300 to 600 mm rainfall.

Linseed (*Linum usitatissimum*)

Linseed is a cool and dry season crop. The temperature ranging between 10 and 25° C is good for growth. High temperature above 32° C accompanied with drought during the flowering stage reduces the yield, the oil content of the seed and the quality of the oil. Linseed is grown as rainfed crop in areas receiving 650 to 1000 mm rainfall. It cannot tolerate heavy rains during growth periods. The plants are susceptible to frost, particularly during flowering.

Caster (*Ricinus communis*)

The castor crop requires a moderately high temperature of 20–27° C with low humidity throughout the growing season. It grows best in areas where there are clear warm sunny days. Prolonged cloudy weather with high temperature above 40° C at the time of flowering results in poor seed set. The plant is considered to be very resistant to drought but even then about 80–100 mm evenly distributed rainfall is required for optimum growth. Heavy rainfall at flowering reduces the yield. Caster plant is very susceptible to frost.

Cotton (*Gossypium* spp.)

The cotton is a warm season crop and is best adapted to subtropical climate. Soil temperature is most important for germination of cotton seed. Minimum, optimum and maximum soil temperature should be 16, 34 and 39° C, respectively. It requires daily atmospheric minimum temperature of 16° C for germination and 21 to 27° C for proper vegetative growth. The crop can tolerate temperatures up to 43° C, but does not do well if the same falls below 21° C. During fruiting phase the day temperature should be 27 to 32° C and nights should be cool.

An annual rainfall of at least 500 mm distributed throughout the growing season is required for a good harvest. Heavy showers and cloudy weather during fruiting period cause shedding of flowers and young bolls. Excessive wet or dry weather reduces boll size. Abundant sunshine during boll maturity and harvesting is essential. Cotton is susceptible to frost.

Jute (*Corchorus* spp.)

Jute plants thrive best under warm and humid climate with temperature ranging from 24 to 37° C, the optimum being around 34° C. A well distributed annual rainfall of 1500 mm is suitable for the crop. Alternate period of sunshine and rainfall is the most beneficial. Young jute plants are sensitive to waterlogging conditions but at later stages of growth they can tolerate waterlogging.

Sugarcane (*Saccharum officinarum*)

Sugarcane is considered as essentially a tropical plant, but can be grown in subtropics. It is a long duration crop and thus it encounters all the seasons, *viz.*, rainy, winter and summer during its life cycle. Growth is closely related to atmospheric temperature. Optimum temperature for sprouting of stem cuttings is 32 to 38° C. It slows down below 25°, reaches plateau between 30 and 34°, is reduced above 35° and practically stops when the temperature is above 38° C. Temperatures above 38° C reduce the rate of photosynthesis and increase respiration. During maturity of canes, however, relatively low temperature of 12 to 14° C is desirable, since this causes the reduction in vegetative growth rate and enrichment of sucrose in the cane. At higher temperatures reversion of sucrose into fructose and glucose may occur besides enhancement of photorespiration thus leading to less accumulation of sugars. Severe cold weather inhibits bud sprouting in ratoon crop and arrests cane growth. Sugarcane cannot tolerate freezing temperatures.

A total rainfall of 1000 to 1500 mm is adequate, provided the distribution is right, abundant in the months of vegetative growth followed by a dry period for ripening. During the active growth period rainfall encourages rapid cane growth, cane elongation and internode formation. But during ripening period high rainfall is not desirable because it leads to poor juice quality, encourages vegetative growth and formation of water shoots. High relative humidity of 80–85% favours rapid cane elongation during grand growth period. However, a moderate relative humidity of 45–65% coupled with limited water supply is favourable during the cane maturity stage. Sugarcane is a sun loving plant. Tillering is affected by intensity and duration of sunshine. High light intensity and long duration promote tillering while cloudy and short days affect it adversely. High wind velocities exceeding 60 km/hour are harmful to grownup canes, since they cause lodging and cane breakage. Also, winds enhance moisture loss from the plants and thus aggravate the ill effects of moisture stress.

Potato (*Solanum tuberosum*)

Climate plays a very important role in potato production. It is a temperate plant and thrives best in cool regions where there is sufficient moisture. Tuber growth is satisfactory when soil temperature is between 17 and 19° C. Higher soil temperature adversely affects the tuber development and above 30° C the tuber development practically stops. At high temperatures, the respiration rate increases, and the carbohydrates produced by photosynthesis are consumed rather than stored in the tuber. It grows best under long day conditions.

Sugarbeet (*Beta vulgaris*)

Sugarbeet is the crop of temperate and cold climate. It can be raised successfully in any region with temperature ranging from 12 to 45° C during the crop season. The favourable soil temperature for germination of sugarbeet seed is 15° C. An average temperature of 20–22° C is ideal for proper growth of the plants and when the temperature is above 30° C sugar accumulations is reduced.

Sweet potato (*Ipomea batatus*)

Moderately warm climate and temperature of 21° to 26° C is very conducive for sweet potato cultivation. It is widely grown in tropical, sub-tropical and warm temperate areas. Temperature below 10° C checks the growth and development of the tubers. A well distributed annual rainfall of 750–1500 mm is favourable for its cultivation. It can tolerate drought to some extent but cannot withstand waterlogging. It requires plenty of sunshine, whereas shade causes yield reduction. High rainfall and long photoperiod encourage vine growth and reduces the tuber yields.

Cabbage (*Brassica oleracea* var. capitata)

In the temperate zones, the cabbages are biennial. For high production the crop requires a cool, humid climate. After sowing, the plants develop vegetative growth and then after exposure to low temperature in winter, they pass into the reproductive phase. The formation of flowers depends on the temperature and not on the day length. A longer period of cold treatment induces the plants to bolt sooner and flowering more abundantly compared to exposure to shorter periods. In the tropics and subtropics, the plants remain vegetative since they do not receive the required cold treatment. Young plants withstand a wider temperature range and they can be grown at higher temperature and planted later in cool season. Optimum growth occurs at a mean daily temperature of about 17° C with daily mean maximum of 24° C and minimum of 10° C. Mean relative humidity should be in the range of 60 to 90%.

Cauliflower (*Brassica oleracea* var. botrytis)

In India, cauliflower is grown in large areas having a cool and moist climate. The cauliflower varieties are very sensitive to temperature. For good seed germination, a temperature of 10–21° C is required. The curding precedes the floral initials and the plant may transform to curding from 5° C to nearly 28–30° C, depending on the cultivar. The optimum temperature required for the growth of young plant is around 23° C and later it drops to 17–20° C. The tropical cultivars show growth even at 35° C. A temperature range of 15–21° C is considered as optimum for growth and curd formation for most of the temperate cultivars of the crop. High temperature during maturity will result in yellowish leafy curds.

Broccoli (*Brassica oleracea* var. italica)

The broccoli is a cool season crop. A temperature in the range of 20 to 25° C is optimum for its proper growth, while the temperature within 15 to 20° C is favoured during heading stage. The heads become loose with rise in temperature. Some cultivars of sprouting broccoli exhibit resistance to frost to a certain extent. These are generally biennial, while the annual cultivars are somewhat sensitive to frost. In such cultivars even a light frost causes considerable damage to the buds.

Carrot (*Daucus carota*)

The carrot is a cool season crop though some of the tropical types are tolerant to quite high temperature. The colour development and growth of the roots are adversely affected by temperature. While a temperature in the range of 7 to 20° C is considered optimum for seed germination, a temperature within 2 to 25° C is ideal for root development.

Radish (*Raphanus sativus*)

The radish is very specific in its climatic requirements especially, temperature. It is predominantly suited to a cool or moderate climate. A high temperature results in bolting before proper root development. Also, the roots become hard, pithy and pungent before reaching maturity. It develops best flavour, texture and size at cooler temperatures between 10 and 15° C.

Brinjal (*Solanum melongena*)

Brinjal or eggplant requires long warm growing season. The optimum soil temperature for seed germination is 24 to 32° C. A long spell of chilly weather may damage the crop. Periods of cool weather during the growing period will retard plant growth and reduce yields. Affected plants seldom recover, even if favourable growing conditions return. Young seedlings are sensitive to frost. The optimum growing temperature range is 21 to 30° C, with a maximum of 35° C and a minimum of 18° C.

Okra (*Abelmoschus esculentus*)

The okra or lady's finger is a tropical crop. Its optimal temperature for germination, growth and fruit setting is between 25 and 30° C. Okra is highly susceptible to frost injury. Seeds fail to germinate when temperature is below 20° C. The plant needs warm weather and plenty of sunshine. The crop gives good yield in warm humid condition. It grows best within a temperature range of 24 to 27° C. It can be successfully grown in rainy season even in heavy rainfall areas.

Lettuce (*Lactuca sativa*)

The lettuce is a cool-weather annual crop and can withstand cold and light frosts, although the degree of tolerance to cold may vary appreciably among cultivars. Heavy frosts will, however, severely scorch the leaves, especially mature heads. The most favourable temperatures for optimum growth and development are daily means between 15 and 18° C, with monthly means between 7 and 24° C. Day temperatures ranging from 17 to 27° C, and night temperatures between 2 and 12° C, are most suitable.

Onion (*Allium cepa*)

The onion is a cool season crop. Growth of onion is affected by length of the day. The requirements of day length of different varieties may differ. Therefore it is better to grow only those varieties which are acclimatised. Temperature is more important than the length of the day in seed-stalk development. Most of the varieties transplanted early (before November in north India) will form seed stalk. The ideal temperature requirement of onion is 12 to 25° C before bulbing and 20 to 25° C for bulb development. A sudden rise in temperature favours early maturity, resulting in small sized bulbs.

Tomato (*Lycopersicon esculentum*)

The tomato is a warm season crop but can be grown intensively in cool season also. The plants cannot tolerate severe frost. The crop does well under an average monthly temperature of 21 to 23° C but commercially it may be grown at temperature ranging from 18 to 27° C. At very high and low temperature there is low germination of seeds, poor plant growth, flower drop, thereby reducing quality and yield of the fruit. Temperature as well as light intensity affects the fruit set, pigmentation and nutritive value of the fruit.

Pumpkin (*Cucurbita moschata*)

The pumpkin is a warm season crop. Pumpkin varieties require 110–120 days to reach maturity. So a frost free season of about four months is essential for their growth. Pumpkins require a mean temperature of approximately 18 to 30° C. High relative humidity and bright sunshine is ideal for its cultivation.

Bitter gourd (*Momordica charantia*)

The bitter gourd is a warm season crop grown mainly in subtropical and hot-arid regions. They are susceptible to light frost. Temperature in the range of 24–27° C is considered as optimum for the growth of the vines. The seed germinates best when temperature is higher than 18° C. High humidity at the time of vegetative growth renders the crop susceptible to various fungal diseases.

Ginger (*Zingiber officinale*)

Ginger is a tropical plant and requires warm humid climate for its cultivation. The base temperature requirement for the crop is 13° C. The upper limit of day temperature is 32° C and night temperature is 27° C, the favourable range being 25–26° C. A rainfall of 1500 to 3000 mm, well distributed in 8–10 months during the growing season is ideal for its cultivation. It is a partially shade loving crop and can be grown in semi-shaded condition. The crop is sensitive to waterlogging and tolerant to drought.

Turmeric (*Curcuma longa*)

The turmeric is a tropical plant. It requires a warm and humid climate. In heavy rainfall tracts of the west coast, it is grown as a rainfed crop and in other areas it is cultivated under irrigation. It can tolerate dry weather conditions and can successfully be cultivated in partial shade. It thrives well at temperature within 20 to 30° C. A well distributed rainfall of 1500 mm or more per annum is good for its cultivation.

Coriander (*Coriandrum sativum*)

The coriander is a tropical crop with a monthly mean temperature requirement of 15 to 25° C. It requires a frost free condition particularly at the times of flowering and seed formation. Dry and moderately cool weather during seed formation increases yield as well as quality of the produce.

Chilli (*Capsicum annuum*)

Chilli prefers a warm humid climate during early stages and a dry weather towards the maturity of pods. It is grown in both tropical and subtropical areas up to 2000 m above mean sea level. As a rainfed crop, it is grown in areas receiving an annual rainfall of 750 to 1000 mm. Excessive or heavy rainfall during its growth causes defoliation and rotting. In north India, chilli is grown in *kharif* and summer seasons as the crop does not withstand the low temperature in winter. But in south India, due to the mild winter chilli is grown in all the 3 seasons. During the growing period which extends over four months, maximum temperature ranging from 20 to 30° C and a minimum temperature not below 10° C is ideal. High temperature during summer accelerates flower and fruit drop causing poor fruit set.

Banana (*Musa* spp.)

Banana is well-suited for cultivation from humid subtropical to semiarid subtropics up to 2000 m above mean sea level. In India, it is successfully grown with a temperature ranging from 15 to 35° C and annual rainfall of 500 to 2000 mm. Its growth may stop below 10° C. If bunch emergence coincides with very low temperature, it results in emergence of inflorescence through pseudostem. Chilling temperature results in similar malformed bunches. Arid zones with strong hot wind and freezing temperature are quite harmful. Photosynthetic rate of banana is drastically reduced in areas with overcast sky. In contrast, excessive light is again harmful which induces bleaching of leaves. Exposed bunches particularly in the uppermost hand and bunch stalk are susceptible to high light intensity. Severe bleaching may cause yellowing of fruit peel which later turns blackish. Tall varieties, in general, are more tolerant to cold temperature above freezing point. High temperature with water stress leads to loss in growth. Water stagnation in poorly drained soils also leads to slow growth. Wind velocity more than 50 km/hour causes serious damage to banana plantation.

Papaya (*Carica papaya*)

Papaya is a plant of tropical climate. However, it also grows well in the mild subtropical region of the country up to 1000 m above mean sea level. It does well in warm and humid climate. Temperature is one of the most important climatic factors which determine the success of papaya cultivation. The optimum temperature for papaya is 22 to 26° C. Low temperature below 10° C affects the growth and development of fruits; it delays maturity and ripening, and the fruit does not develop flavour causing poor taste. High temperature tends to induce more maleness in papaya plant that produces flowers with varying degrees of maleness and femaleness. It cannot tolerate frost, strong wind and water stagnation. It requires an annual rainfall of 1500 to 1800 mm distributed equally. Comparatively high temperature, low humidity and adequate availability of soil moisture are conducive for better quality fruits.

Orange (*Citrus sinensis*)

The oranges prefer more subtropical humid climate with high rainfall. Orange trees are climate sensitive plants that have quite definite temperature requirements. The crops can be cultivated successfully in the tropical and subtropical regions with rainfall ranging from 850 to 3000 mm uniformly distributed from March to November and temperature ranging from 10 to 35° C. High winds and hailstorm during flowering and fruit set are highly detrimental for fruit bearing.

Guava (*Psidium guajava*)

The guava is adapted to tropical and subtropical climate. The optimum temperature required for cultivation is between 23 and 28° C. However, best quality guava is obtained where low night temperature of 10° C prevails during winter months. It can tolerate high temperature and drought conditions but it is susceptible to severe frost as it can kill young plants. An annual rainfall of 1000 to 2000 mm is ideal for guava cultivation. Occurrence of rain during harvesting period deteriorates the quality of fruits.

Litchi (*Litchi chinensis*)

Litchi grows well in moist subtropical climate having abundant rainfall. The frost in winter and dry heat in summer are the limiting factors for litchi cultivation. It grows luxuriantly at 30° C. The maximum temperature prevalent during flowering and fruit development varies from 21° C in February to 38° C in June. Prolonged rain at the time of flowering is not desirable. Humidity is another critical factor for litchi. The dry winds cause fruit cracking and subsequent damage of the pulp. Wet spring, dry summer and light winter are desirable conditions for fruiting in litchi.

Pine apple (*Ananas comosus*)

Pineapple grows in warm and humid climate. The optimum temperature is from 15° to 32°C for normal growth. High temperature over 35° C is unfavourable for the development of fruits, especially if the relative humidity is low. Exposure of the fruits to strong sunshine leads to sun scalding. It can be grown up to an elevation of 1100 m above the sea level, if these places are free from frost, have a relatively high atmospheric humidity with an average rainfall of 760–1000 mm.

Coconut (*Cocos nucifera*)

The coconut palm is essentially a tropical plant. The palm requires an equitable warm and humid climate neither very hot, nor very cold. The mean annual temperature for optimum growth and maximum yield is stated to be 27° C with a diurnal variation of 6 to 7° C. Annual rainfall of 1000 to 2000 mm, well distributed throughout, is ideal for proper growth and maximum production.

Rose (*Rosa* spp.)

Rose plants can survive in wide range of temperature but the ideal average temperature for optimum cut-flower production lies in the range of 24 to 28° C at day and 18 to 20° C at night. Rose grows well under relative humidity range of 65 – 70%. Light intensity between 60000 and 70000 LUX supports production of quality flowers.

CLIMATIC NORMALS FOR LIVESTOCK PRODUCTION

Weather and climate have also great influence on farm animal production. The environment in which an animal lives is referred to as its habitat. A habitat includes both biotic (living) and abiotic (non-living) components of the animals' environment. Abiotic components of an animal's environment include temperature, humidity, oxygen, wind, soil composition, day length and elevation. Knowledge of how potential environmental stressors (ambient temperature, humidity, thermal radiation, air speed, *etc.*) can directly and adversely affect animal performance, health and well-being is also required for a successful animal farming. The indirect consequences of weather include their impact on feed quality and availability.

Birds and mammals are warm-blooded (endotherms or homeotherms), possessing a thermoregulatory system that maintains a stable body temperature and uniform internal environment (homeostasis) by converting food energy into heat. All metabolic processes generate substantial heat and much of this is retained within the body of homeotherms by insulating layers of fat, feathers or fur. Dairy cattle, like other warm-blooded animals, function most efficiently in environments where they can maintain their body temperature at around 38° C. Tissue and cellular metabolism and the underlying biochemical reactions that sustain life and productive functions need body temperature to be maintained within very narrow limits. Relatively small increases in body temperature result in the breakdown of body protein and a significant depression in production.

Effect of heat stress on cattle

All animals have a range of ambient environmental temperatures termed as the thermo neutral zone. This is the range of temperatures that are conducive to health and performance. The upper critical temperature is the point at which heat or thermal stress effects begin to affect the animal. The environmental factors that contribute to heat stress include high temperature, high humidity and sunlight. Heat stress can be simply defined as the point where the cow cannot dissipate an adequate quantity of heat to maintain body thermal balance.

The environmental conditions that induce heat stress can be calculated using the temperature humidity index (THI). A THI is a single value representing the combination of air temperature and humidity associated with the level of thermal stress. The THI was developed during early 1990s as a weather safety index to monitor and reduce heat stress-related losses. Utilising both humidity and temperature is important compared with utilising temperature or humidity alone to evaluate heat stress. THI is calculated from air temperature and relative humidity using the following equation.

THI = Dry bulb temperature in degree Celsius + (0.36 × dew point temperature in degree Celsius) + 41.2

The dew point is the temperature at which the water vapour in a sample of air at constant barometric pressure condenses into liquid water at the same rate at which it evaporates. Heat stress begins to occur in dairy cattle when the THI is > 72. The THI threshold values for beef cattle, swine and poultry are 72–75, 72–74 and 70–78, respectively.

Table 12.4 Effect of heat stress on dairy cattle based on temperature humidity index

THI	Stress level	Comments
< 72	None	□
72–79	Mild	Dairy cows will adjust by seeking shade, increasing respiration rate and dilation of the blood vessels. The effect on milk production will be minimal.
80–89	Moderate	Both saliva production and respiration rate will increase. Feed intake may be depressed and water consumption will increase. There will be an increase in body temperature. Milk reduction and reproduction will be decreased.
90–98	Severe	Cows will become very uncomfortable due to high body temperature, rapid respiration and excessive saliva production. Milk production and reproduction will be markedly decreased.
> 98	Danger	Death of cow may occur

(*Source*: Hahn, 1999 and West, 2003)

Impacts of thermal stress on animal health

Thermal stresses have both direct and indirect effects on animal health. Direct effects of thermal stress include simple physiological disturbances, organ dysfunction and death. Reduction in feed intake together with diversion of more energy to maintain normal body function creates negative energy balance which deteriorates animal health conditions. These animals with poor health are prone to infectious diseases. The high ambient temperature and moisture level creates the environment suitable for fungus growth in feed and feedstuffs which may lead to mycotoxicosis in animals due to mycotoxin produced by fungi.

Impacts of thermal stress on reproduction

Thermal stress causes imbalance in secretion of reproductive hormones. High ambient temperature increases incidence of ovarian cysts and reduces fertility of cattle. Conception rates are less in cattle under thermal stress as compared to those in thermo-neutral zone. Thermal stress prior to and immediately after artificial insemination causes reduction in conception rate in high producing lactating cows. Similarly, climatic warming also affects reproductive performance in bulls.

Impacts of thermal stress on milk production

Reduction in milk production is one of the major economic impacts of climatic stress in dairy cattle. Reduction in milk yield is further intensified by decrease in feed consumption by the animals. Reduced milk production due to thermal stress is attributable only partly to decrease in feed intake. Actually 35% of reduced milk production is due to decreased feed intake while remaining 65% is attributable to direct effect of thermal stress (Bajagai, 2011). Milk production in cow has been found to be reduced when ambient temperature and temperature humidity index increase above critical threshold. The yield and composition of milk are affected by the temperature of the environment. Environmental temperatures lower and higher than the comfort zone temperatures reduce yield and alter many of the components of milk. At all India level an estimated annual loss due to direct thermal stress on livestock is about 1.8 million tonnes of milk, that is, nearly 2% of the total milk production in the country. This decrease can be either transitory or long term depending on the length and severity of heat stress. The decline in milk production will be higher in crossbreds followed by buffalo and indigenous cattle.

EFFECT OF CLIMATE ON POULTRY BIRDS

Hot weather can have a severe impact on performance of poultry. The climate in poultry houses influences the health of the birds. Respiratory, digestive and behavioural disorders are more likely to occur in houses in which the climatic conditions are not up to standard. Production efficiency can be affected long before the temperature reaches a level at which survival becomes a concern.

Table 12.5 Effect of ambient temperature on adult poultry birds

Temperature range (in degree Celsius)	Effect on adult poultry birds
13–24	Thermal neutral zone. The temperature range in which the bird does not need to alter its basic metabolic rate or behaviour to maintain its body temperature.
19–24	Ideal temperature range.
24–30	A slight reduction in feed consumption can be expected, but if nutrient intake is adequate, production efficiency is good. Egg size may be reduced and shell quality may suffer as temperatures reach the top of this range.
30–33	Feed consumption falls further. Weight gains are lower. Egg size and shell quality deteriorate. Egg production usually suffers. Cooling procedures should be started before this temperature range is reached.

Temperature range (in degree Celsius)	Effect on adult poultry birds
33–35	Feed consumption continues to drop. There is some danger of heat prostration among layers, especially the heavier birds and those in full production. At these temperatures, cooling procedures must be carried out.
35–38	Heat prostration is probable. Emergency measures may be needed. Egg production and feed consumption are severely reduced. Water consumption is very high.
> 38	Emergency measures are needed to cool birds. Survival is the concern at these temperatures.

(*Source*: Anderson and Carter, 2007)

The layers are warm blooded with a body temperature of 41–42.2° C. Body temperature is kept quite constant and is regulated by part of the chicken brain. It takes some time before heat regulating mechanisms start functioning in newborn chicks and therefore they need a higher ambient temperature than adult birds do. Furthermore, the ratio between the surface area and weight of young chicks is unfavourable and they do not have any fat reserves. The comfort zone is defined as the temperature zone in which the birds are able to keep their body temperature constant with minimum effort. Behaviour of birds will change when temperatures rise to above or decrease to below the comfort zone.

The thermo-neutral zone is defined as the temperature zone in which the birds are able to keep their body temperature constant with the help of physical heat regulation. The lowest temperature in the thermo-neutral zone is called the lowest critical temperature. If temperatures fall to under this temperature the bird will start to use feed energy to warm itself and will consequently consume more feed. The highest temperature in the thermo-neutral zone is called the highest critical temperature. If the temperature rises above this temperature the birds can no longer dissipate their heat. They will start to consume less feed and production will drop as a result. The highest and lowest critical temperature of a bird depends on its age and bodyweight, housing system, feeding level, relative humidity and air velocity. The critical temperature for layers is 20° C. For every 1° C lower than 20° C, the birds require an extra 1.5 g of feed per day. The most efficient temperature for layers lies between 20 and 24° C. When temperature rises above 24° C, shell quality and egg weight will reduce. The critical temperature for broilers and rearing birds is highly dependent on age.

Relative humidity in poultry houses is measured to determine its impact on respiratory disorders. If the relative humidity is too high, condensation may accumulate in the house which leads to wet litter. This has a direct effect on the growth of microorganisms. Humidity is controlled by the intense heating or cooling of air inside the house in response to the temperature outside the house. The ideal relative humidity for poultry is 60–80%.

Table 12.6 Recommended temperature and relative humidity for poultry birds

Age of the bird (in days)	Recommended ambient temperature (°C)	Recommended relative humidity (%)
1	33	50
7	31.5	50
14	28.5	60
21	28.5	65
28	26	70
42	23	85

(*Source*: Hulzebosch, 2006)

Optimum temperature for ducks

At the time of hatching, ducklings require a high temperature of about 30° C. They are not yet able to regulate their body temperature and require supplemental heat such as that provided by a brooder. As they grow older they become better able to produce and conserve heat, and regulate their body temperature. After a duckling is fully covered with feathers, they are able to maintain proper body temperature even when the outside temperature is low.

Table 12.7 Recommended temperatures for ducks

Age of ducks in days	Temperature in °C
1	30
7	27
14	23
21	19
28	15
35	13
42	13
49	13
Developing breeders	13
Laying breeders	13

(*Source*: Scott and Dean, 1991)

MODIFICATIONS OF CROP MICROCLIMATE

Any climatic condition in a relatively small area, within a few metres or less above and below the earth's surface and within canopies of vegetation is known as microclimate. The term usually applies to the surfaces of terrestrial and glaciated environments, but it could also pertain to the surfaces of oceans and other bodies of water. The scale of climate can be viewed in three dimensions such as time, horizontal distance and vertical height above the ground. These dimensions are defined by the World Meteorological Organization.

Table 12.8 Climate scales and their dimensions

Scale name	Time dimension	Horizontal dimension	Vertical dimension
Microclimate	1 second to 1 hour	1 mm to 1 km	1 mm to 10 m
Mesoclimate	1 hour to half day	1 km to 100 km	10 m to 1 km
Macroclimate	Half day to 1 week	100 km to 10,000 km	1 km to 20 km
Global climate	1 week and longer	> 10,000 km	20 km to 100 km

Microclimatic conditions depend on environmental factors such as temperature, humidity, light, wind speed, dew, frost, heat balance, and evaporation. The soil type also has a considerable effect on microclimate. Sandy soils and other coarse, loose, and dry soils are subject to high maximum and low minimum surface temperatures. Soils of lighter colour reflect more and respond less to daily heating. The ability to absorb and retain moisture of soil is influenced by its composition. Presence of vegetation and its type has also considerable impact on the microclimate of a place as it controls the flux of water vapour into the air through transpiration. In addition, vegetation can insulate the soil below and reduce temperature variability.

Topography can affect the vertical path of air in a locale and, therefore, the relative humidity and air circulation. Air ascending a mountain undergoes a decrease in pressure and often releases moisture in the form of rain or snow. As the air proceeds down the leeward side of the mountain, it is compressed and heated, thus promoting drier, hotter conditions there. An undulating landscape can also produce microclimatic variety through the air motions produced by differences in density.

The microclimates of a region are defined by the moisture, temperature, and winds of the atmosphere near the ground, the vegetation, soil, and the latitude, elevation, and season. Weather is also influenced by microclimatic conditions. Wet ground promotes evaporation and increases atmospheric humidity. The drying of bare soil, on the other hand, creates a surface crust that inhibits ground moisture from diffusing upward, which promotes the persistence of the dry atmosphere. Microclimates control evaporation and transpiration from surfaces and influence precipitation, and so are important to the hydrologic cycle.

The growth of every crop is influenced by various weather parameters like temperature, radiation, sunshine, rainfall, humidity and wind speed. Any deviation in these parameters affects the normal growth of the crop. Therefore, excesses and deficits cause great stresses. Each component of the microclimatic environment exhibits unique spatial and temporal responses to its relationships with the biological processes. These relationships are complex and often non-linear. These features are well visualized with the facts that temperature, solar radiation and humidity affect plant growth by influencing physiological processes such as photosynthesis, respiration, seed germination, mortality and many enzymatic activities. The ecosystem processes such as decomposition, nutrient cycling, succession and productivity are partially dependent on microclimatic variables. Many animals are also adapted to specific microclimatic conditions. Wind speed, air temperature, humidity, and solar radiation can influence migration and dispersal of flying insects. Soil microbe activity is affected by soil temperature and moisture. Many fish species have specific thermal ranges in which they are able to survive and reproduce.

Artificial control of field environment to keep the optimum condition of plant growth and crop production is known as crop microclimate or field climate modifications. A sound knowledge on physiology of crop plants and their physical environment is required for crop microclimate modifications. This can be done mainly through controlling wind velocity, heat load, radiation balance and water balance.

Radiation can be manipulated by mutual shading by the crop plants, increase or decrease of surface absorption and cover to minimize radiation loss at night. When plants begin to grow, the microclimate of the field is modified. In a short period of time, the leaves of one plant begin to touch the leaves of other adjacent plants. These plants and leaves tend to interfere with exchange of heat, moisture and momentum between ground and the atmosphere. When their leaves start shading the ground completely, top of the crop canopy becomes active surface for heat and other exchanges and the soil surface becomes a secondary one. Transpiration and thermal radiation from the plant parts within the crop canopy constitutes a tertiary source for the energy and moisture fluxes.

Every crop has the tendency to develop its own stand and form a microclimate with different characteristics. During heat exchange in and over a vegetative surface, the plant leaves taking part in various forms of disposal of absorbed radiation has very small thermal capacities. The plant parts cast their shadows on the soil surface which decreases the heat exchange in the soil between the soil and crop air layer. Thus, the flow of heat entering or leaving the soil and the leaves on it and the air layer in and below the canopy is very small. Reduced transpiration due to soil water shortage during day time push up the leaf temperature by 5-10° C above that of the air. The desired microclimate may be achieved by adjusting or altering the plant population, planting geometry, spacing, planting time and sometimes crop mixtures.

Excessive rainfall in any area causes adverse effect on the crop growth. Similarly, moisture deficit also causes stress by affecting the exchange processes. Extreme temperature conditions are harmful for the crops. Low temperature conditions during winter season and high temperature conditions during summer season severely affect the crops. Heat and/or moisture flow in a crop field can be manipulated by agronomic practices like no-tillage or zero-tillage, mulching, crop residue management, application of soil amendments and efficient crop rotation.

13

Weather Forecasting

Weather forecasting is the application of science and technology to predict the state of the atmosphere for a given location. Weather forecasts are made by collecting quantitative data about the past and current state of the atmosphere on a given place and using scientific understanding of atmospheric processes to project how the atmosphere will evolve on that place. Weather warnings are important forecasts because they are used to protect life and property. Forecasts based on temperature and precipitation is important to agriculture.

Agricultural production is highly dependent on weather, climate and water availability, and is adversely affected by weather and climate related disasters. In many developing countries where rainfed agriculture is the norm, a good rainy season means good crop production, enhanced food security and a healthy economy. Failure of rains and occurrence of natural disasters such as floods and droughts can lead to crop failures, food insecurity, famine, loss of property and life, mass migration, and negative national economic growth (Sivakumar, 2006).

Farming in many parts of the world, especially in the arid and semiarid regions is risky, because climate is highly variable. In some developing countries as much as 80% of the variability in agricultural production is due to the variability in weather conditions. This production variability in agriculture has immediate and important macroeconomic impacts. A timely seasonal forecast of favourable conditions could permit farmers to adjust cropping patterns and input use in order to benefit fully from these conditions. Also, a timely forecast could give the marketing system and downstream users, such as the food processing industry, time to prepare for a bountiful harvest (Arndt *et al.*, 2000). It is important to remember that under rainfed conditions, even a less reliable, but earlier forecast may be more valuable than an accurate but late forecast (Mjelde *et al.*, 1998).

In about 340 B.C., Aristotle described weather patterns in *Meteorologica*. Later, Theophrastus compiled a book on weather forecasting, called the *Book of Signs*. Ancient weather forecasting methods usually relied on observed patterns of events, also termed pattern recognition. This experience was accumulated over the generations to produce weather knowledge. However, not all of these predictions prove reliable, and many of them have since been found not to stand up to rigorous statistical testing.

Two officers from of British Royal Navy, Francis Beaufort and Robert FitzRoy formed the basis for all of today's weather forecasting knowledge by middle of 19th century. A terrible storm in 1859 inspired FitzRoy to develop charts to allow predictions to be made, which he called 'forecasting the weather', thus coining the term 'weather forecast'. The 'Weather Book' which FitzRoy published in 1863 was far in advance of the scientific opinion of the time.

With the invention of the telegraph and telephone, the study of weather systems took off. Scientists could then study masses of weather and how such large systems could move and in what direction. By the early 20th century, scientists had begun to believe that weather forecasting could be predicted based on mathematical methods. Computers were first used to create surface analyses in 1992 and by 1998 digital imagery were used for satellite, radar and surface observations, encompassing nearly every factor of weather interpretation. Modern technology, particularly computers and weather satellites, and the availability of data provided by coordinated meteorological observing networks, has resulted in enormous improvements in the accuracy of weather forecasting.

METEOROLOGICAL AND CLIMATOLOGICAL FORECASTS

The meteorological forecasts are defined as forecasts of the state of the atmosphere, particularly the near surface atmosphere, from the present up to 15 days into the future. These forecasts may be contrasted with climatological forecasts which are defined as a forecast of the mean atmospheric conditions more than 15 days into the future (Hollinger, 2009).

The major differences between the two forecasts are the lead time and the length of time for which the forecast represents. Lead time is defined as the duration of gap between the time of delivery of forecast and the effective time of occurrence of the forecast event. Meteorological forecasts are made hours to days before the effective time of the forecast, while climatological forecasts are made months to years before the effective time of the forecast. Meteorological forecasts provide estimates of actual temperature; humidity; precipitation probability, type and amount; wind speed and direction; and sky conditions. However, climatological forecasts are given in terms of deviation from normal, such as above normal, normal, or below normal. Thus, the forecast is incomplete without the knowledge of the normal temperature and precipitation conditions for the time and location of the forecast. Normals are defined by thirty years of daily weather observations, and are updated every 10 years.

Meteorological and climatological forecasts can be 'general' or 'applied'. A general forecast represents the conditions without reference to a particular activity, while an applied forecast takes the general forecast and applies it to an activity such as agricultural and animal production, or aviation. Meteorological and climatological forecasts for agricultural production differ from general meteorological and climatological forecasts in their application. For agricultural production the general forecasts are interpreted to indicate how the forecast conditions may impact the growth and development of crops, diseases, insect pests, and livestock; and how the conditions will affect production management practices. Thus, meteorological and climatological forecasts for agricultural production require knowledge of the circulation of the atmosphere and the factors that cause the day-to-day changes in the weather, and how agricultural organisms respond to short and long term atmospheric conditions. Crop canopies develop a microclimate and it is within these conditions that the crop interacts with the soil, pests, and diseases. Therefore, the micrometeorological conditions must be related to the synoptic-scale meteorological forecast (Hollinger, 2009).

The meteorological or climatological forecast is to be provided with a reasonable degree of accuracy and enough lead time so that a farmer can take the necessary actions to protect his crop or livestock. He must be able to change his management practices to take advantage of favourable weather conditions, or mitigate u. favourable conditions. A perfect forecast has the potential to have the most value, or even an imperfect forecast can have some value if the forecast is in the right direction and provided with enough lead time for management practices to be effected. However, a highly accurate forecast may have no value if appropriate management practices could not be applied to the crop or livestock before the forecast event.

TYPES OF WEATHER FORECASTING

The product or result of a weather forecast can be expressed in a variety of forms, which are called type of forecast. The forecast may be of quantitative, qualitative and probabilistic in nature. Depending on the spatial and temporal scales of atmospheric systems and the details of the accuracy desired, the weather forecasts are usually referred to nowcasting, short range, medium range and long range.

Nowcasting

Prediction on the time scale of minutes to hours is called nowcasting. This type of weather forecasting is issued for less than 12 hours. It is concerned with severe weather events such as tornadoes, hail, and flash floods, which are manifestations of intense mesoscale weather systems. Nowcasting techniques use highly automated computers and image analysis systems to integrate data rapidly from a variety of sources.

Short range forecasting

It comprises of forecast and warning of weather elements which is valid for 24 hours and an outlook for the subsequent two days. It includes the day-to-day prediction, usually concerned with the more ordinary weather events and changes in the synoptic-scale weather systems. The short range forecast includes cloud spread, rainfall distribution, heavy rainfall warning, maximum and minimum temperature, relative humidity, duration of sunshine, heat and cold waves, low pressure areas, cyclone warning, hail or thunderstorm, snow, frost and likelihood of maximum wind speed. The information is sufficiently accurate. Though it helps in scheduling irrigation, adjusting time of agricultural operations and protecting plants from frost, this type of forecast is not much beneficial to the farmers because the lead time of the forecast is too short to take preventive measures against the adverse weather effects. The short range forecast is issued twice daily. This forecast information is available through television, radio and newspapers. It may also be obtained *via* the telephone from the forecasting agencies. A short range forecasting is based on a detailed analysis of the physical processes occurring in the atmosphere.

Medium range forecasting

It is a forecast and warning of weather elements which is valid for 3 to 10 days. It is also known as extended forecast. This forecast includes cloud amount, rainfall distribution, sequence of rainy days and weekly cumulative rainfall, extended dry or wet spells, maximum and minimum temperature, and average wind speed. The forecast information is sufficiently accurate. The extended weather forecast is most effective and useful in agriculture as it gives sufficient lead time for both planning and executing farm operations. It helps to determine sowing time and depth of sowing, in planning of irrigation schedules, in making decisions on harvesting and time of spraying to get higher efficiency and in managing labourers and equipments. The forecasting procedure is fundamentally same as for short range forecasting. This method involves the production of a number of forecasts using the same computer model but slightly altering the initial conditions within the error range of the observational instruments. It involves enormous numerical computations. Medium range forecasting is receiving considerable attention worldwide, particularly in agriculture sector. In India, the National Centre for Medium Range Weather Forecasting (NCMRWF) has been established in New Delhi during 1988, to develop only medium range weather forecasting for 3 to 10 days ahead. At present the NCMRWF is giving the forecast along with agroadvisories twice a week for all the 127 NARP agroclimatic zones of the country.

Long range forecasting

The forecast valid for more than 10 days is knows as long range forecasting. There is no rigid definition for long range forecasting. Generally the forecast period ranges from a week to the limits of technical feasibility. The India Meteorological Department (IMD) started issuing long range forecasting since 1988 on total monsoon rainfall of the country by 25th May every year. It has emphasised on abnormality of temperature and precipitation. The seasonal predictions can be described as above normal, near normal or below normal. This forecast will be helpful in deciding soil moisture management, irrigation scheduling, selection of crops, managing irrigation with limited water supply and deciding cropping pattern and crop yield. These forecasts can be used for predicting likely trends in agricultural production before the beginning of the *kharif* season. The methods used for long range forecasting include climatology, statistics, numerical models and subjective judgement.

Climate/agroclimate forecasting

The trends in rainfall, its variability and probability of distribution over a season can be determined weekly by using the historical data on rainfall for a good number of years at a particular location. This information is useful to the farmers as crop growth periods can be adjusted depending upon the rainfall probability. This forecasting predicts the impact of climatic variability on the agricultural production over a period of time. By analysing and computing the past data on rainfall and evapotranspiration of a locality the periods of water surplus and deficit can be predicted. This information will be helpful in selecting crops and adjusting their growth periods, irrigation scheduling during the dry spells, and facilitating proper drainage during water surplus. The past biological data of crops in a specific location can be interrelated to the meteorological data pertaining to the period and location. This information will help in adopting suitable agronomic management practices that can mitigate the adverse effects of weather aberrations. There are some fundamental differences in predicting weather and climate. Weather prediction is largely restricted to the consideration of dynamic and physical processes internal to the atmosphere. In climate prediction interactions between the atmosphere and other elements of the climate system are taken into account.

TOOLS OF WEATHER FORECASTING

The weather forecasting requires several basic inputs such as area coverage under the network of meteorological stations, collection of meteorological data on cloud cover, sunshine, radiation, rainfall, temperature, wind and pressure, and upper air data on wind, humidity and temperature at various heights. It requires an international effort even while making forecast for a small region. The World Meteorological Organisation (WMO) under the auspices of the United Nations oversees the collection of weather data across the globe from its 179 member nations. The WMO collects data through its global observing system called World Weather Watch (WWW) which comprises of 10,000 land stations, some 7,000 ship stations and 600 moored and drifting buoys with automatic weather sensors. The weather data thus collected at different observation stations from all countries are sent to the three world meteorological centres at Washington DC, Moscow and Melbourne, which in turn disseminate the data to all member countries. The member nations of the WMO maintain their own meteorological agencies that obtain and process the data and issue regional and national forecasts. The weather forecasting agencies use various tools for collection of the weather data, coding and plotting of these voluminous data to prepare charts. They also need information on the state and stage of crops to give weather forewarnings. The crop weather calendar is required to deliver weather services like forecasting with agroadvisories.

Pilot balloons

The pilot balloons are small balloons usually made of a highly flexible latex material inflated with a gas lighter than air, *i.e.*, hydrogen or helium. These are generally launched to determine wind speed and wind direction at different heights in the atmosphere using a theodolite. Sometimes they may also be used for a wide variety of observations. The theodolite is an optical telescope that measures horizontal and vertical angles of the ascending balloons over a specific time interval. High accuracy is obtained when using a dual theodolite set-up. The ascent rate can be controlled by the amount of gas with which the balloon is filled. The path of the balloon is determined from the data observed through graphical techniques.

The balloons are provided in different colours to provide better sighting depending on the sky background. White balloons are used when there is a clear sky, black balloons are used when there is a low overcast and red balloons are used when there is high overcast, or there is a white or grey background. The radio-theodolites are also used to track the balloon when the sky is obscured.

Radiosonde

Radiosondes are small lightweight boxes equipped with weather instruments and a radio transmitter attached to the ascending balloons. As the radiosonde rises at about 300 metres/minute, sensors on the radiosonde measure profiles of pressure, temperature, and relative humidity. These sensors are linked to a battery powered 300 milliwatt or less radio transmitter that sends the sensor measurements to a sensitive ground tracking antenna on a radio frequency. The meteorological instruments contained in the radiosonde are designed to register temperature ranging from about +40 to –90° C, relative humidity from 15 to 100% and barometric pressure from 1040 to 40 mb or less. The radio signals received by the tracking antenna are converted to meteorological values and from these data significant levels are selected by a computer, put into a special code form, and then transmitted to data users.

The advantage of radiosonde is that the data are available very fast with it. Usually a complete radiosonde ascent takes about two hours and can ascend to over 35 km and drift more than 300 km from the release point. The radiosonde is suspended 25 to 35 metres below the balloon to minimise contamination of the temperature measurements from heat shedding off the balloon skin. A radiosonde that is dropped from an airplane and falls, rather than being carried by a balloon is called a dropsonde. Balloons meant to stay at a constant altitude for long periods of time are known as transosondes. Specialised radiosondes are used for measuring particular parameters, such as the ozone concentration. Radiosondes are essential sources of meteorological data, and hundreds of them are launched all over the world daily.

Weather satellites

Weather forecasting has been greatly benefited from major advances in remote sensing. For viewing large weather systems on a worldwide scale, weather satellites are invaluable. Geostationary and polar orbiting satellites show cloud formations, large weather events such as hurricanes, and other global weather systems.

There are two types of sensors. One is a visible light sensor called the 'imager', which works like a camera in space and helps gather information on cloud movements and patterns. This sensor can only be used during daylight hours, since it works by capturing reflected light to create images. Since different surface features reflect light in distinctive ways, they can be distinguished from each other in the images. Water reflects very little light, making it appear black on the satellite image. Land masses tend to appear as grey shades, depending on their temperature and moisture. The second sensor is called the 'sounder'. It is an infrared sensor that reads temperature. The higher the temperature of the object, the more energy it emits. This sensor allows satellites to measure the amount of energy radiated by earth's surface, clouds, oceans, air, and so on. Infrared sensors can be used at night, a helpful feature for forecasters, considering that the imager can only pick up data during daylight hours.

In addition to providing visual images, satellites can also provide data that allow calculation of atmospheric temperature and moisture profiles and other environmental variables. This is done using a variety of instruments, including atmospheric sounders, which measure quantities at various levels in atmospheric columns. The data retrieved from sounder measurements taken from a satellite can be utilised in a way similar to radiosonde observations, with the major advantage that the satellite data are more complete spatially, filling in gaps between weather ground stations, which often are hundreds or even thousands of kilometres apart.

There are very few upper air stations over the oceans that cover 70% of the earth's surface. Satellites are now providing some observations of wind and temperature profiles over the oceans. However, the signals received by the satellites are affected by the distribution of humidity and cloud cover. Therefore, information on wind and temperature derived from satellites suffer from a certain amount of uncertainty. Nevertheless, due to their large spatial coverage, this information is valuable in making weather forecasts.

Radar

Doppler radar is another essential meteorological tool. Radar is an acronym that stands for Radio Detection and Ranging. It works a little differently from satellite sensors. Instead of reading reflected light or energy, radar measures reflected sound waves. When sound waves are broadcast from a radar antenna, they may come into contact with objects in their path, such as dust particles or ice crystals. If they come into contact with an object that is moving away from the radar, the sound waves will be reflected back at a decreased frequency. If the object they come in contact with is moving towards the radar, the sound waves will be reflected back at an increased frequency. This effect was discovered in 1842 by Christian Doppler. Later, scientists applied Doppler's principle to weather radar. Using Doppler radar, meteorologists can get a picture of precipitation that allows them to track a storm's progress over time. The radar is used for detecting tropical cyclones and their movements. The IMD has been operating radar systems in the east and west coasts of the country.

Some radiosondes are tracked by radar as they ascend through the atmosphere, to determine the wind speed and direction in the middle and upper atmosphere. Radiosondes tracked by radar are called rawinsondes. Over the whole globe there are about 1100 such stations but only about 600 of these provide regular observations twice a day at noon and midnight Greenwich Mean Time.

Upper level information is obtained from other sources as well. Many commercial aircrafts are equipped with weather sensors that automatically monitor the atmosphere throughout their flights. Weather satellites supplement the upper level database by determining temperature and humidity at several depths throughout the atmosphere. Aircrafts and satellites provide data from locations far away from any radiosonde site, particularly over the oceans.

Weather maps

An enormous volume of meteorological data is being collected from all over the world, continuously round the clock, through various telecommunication channels. To assess, assimilate and analyse the vast data, they have to be suitably presented. For this purpose, the observations are plotted on maps using standard weather codes (table 13.1). These maps are called synoptic maps or charts and are the most valuable tool for weather forecasting. Weather forecasting requires analysing the atmospheric condition both at the surface and upper air. Hence meteorologists use two types of synoptic maps namely the surface maps and upper-level maps.

Table 13.1 Some of the codes used on weather charts

Codes	Weather parameters
TTT	Dry bulb temperature in degree Celsius
TdTdTd	Dew point temperature in degree Celsius (plotted in red colour)
TxTxTx	Maximum temperature
TnTnTn	Minimum temperature
ww	Present weather
VV	Visibility
PPP	Barometric pressure reduced to sea level
ddd ff	Wind direction and speed
C_L	Types of low cloud
C_M	Types of medium clouds
C_H	Types of high clouds (plotted in red colour)
N	Total amount of cloud
h	Height of the base of low cloud
Nh	Amount of cloud reported under h
RRR	Amount of rainfall in mm (plotted in green colour)

Synoptic charts display the weather conditions at a specified time over a large geographical area. The surface synoptic charts plotted for different synoptic hours depict the distribution of pressure, temperature, dew point, clouds and winds. The upper air charts are prepared at the 850, 700, 500, 300 and 200 mb pressure levels of the atmosphere, which correspond to nearly 1.5, 3, 5.6, 9 and 12 km above mean sea level, respectively. The pressure, wind and temperature data are plotted at each level. The surface charts together with the upper air charts provide a composite three dimensional weather picture pertaining to a given time. The pattern of pressure distribution is brought out by drawing isobars, troughs, ridges, lows, highs, depressions, cyclones, fronts and discontinuities. Charts are updated at least every six hours. The meteorological observations are plotted according to standard station models prescribed by WMO. The synoptic charts together with the satellite pictures on cloud spread are the basic material needed to understand the initial state of the atmosphere. These systems are clearly marked and labelled by using appropriate symbols (table 13.2).

Table 13.2 Standard symbols used on weather charts

Symbol	Precipitation	Symbol	Cloud cover	Symbol	Wind speed
	Drizzle		Clear sky		Calm
	Shower		*One okta*		1-2 knots
	Rain		Two oktas	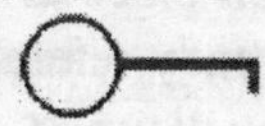	5 knots
	Snow		Three oktas		10 knots
	Hail		Four oktas	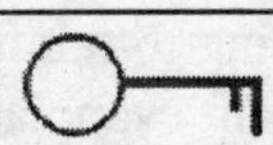	15 knots
	Thunder-storm		Five oktas	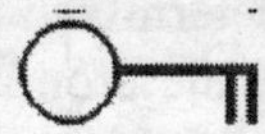	20 knots
	Heavy rain		Six oktas		50 knots or more
	Sleet		Seven oktas		
	Snow shower		Eight oktas		

Symbol	Precipitation	Symbol	Cloud cover	Symbol	Wind speed
=	Mist	⊗	Sky obscured		
≡	Fog				

Crop weather calendar

In order to provide the farmers with an efficient weather service, it is essential that the weather forecaster should be familiar with the crops that are grown in a particular agroclimatic zone. The type of forewarnings to be given depends on the stages of the crop. The users of the forecasts, *viz*., farmers should become familiar with weather based agro-bulletins and learn how to interpret them. To meet the above requirement, the detailed information collected from the agricultural departments has been condensed by the IMD and presented in a pictorial form known as crop weather calendar.

The crop weather calendars consist of three parts, *viz*., crop husbandry, normal weather requirements and weather warnings. At the bottom is the typical life history of the crop, from the sowing window to the period of maturity, in the form of a diagram. Important crop phases like sowing, germination and seedling, transplanting, vegetative growth, flowering, grain formation and maturity are indicated against the standard agrometeorological weeks. The phenological stages of the crop cover certain time intervals indicated by horizontal arrows, which depend on variations in crop variety, sowing date from place to place and from year to year, and the nature of the crop itself. For ready reference, the months and standard weeks are marked at the bottom of the diagram.

Middle part of the calendar gives information regarding normal weather condition required for active crop growth. It is divided into different sections according to rainfall, rainy days, minimum temperature, maximum temperature, pan evaporation and sunshine hours. The normal weather conditions are determined from long term averages. Normal phasewise water requirements of the crop are determined from potential (or reference crop) evaporation multiplied by appropriate crop coefficients. The uppermost portion of the calendar indicates the weather conditions favourable for incidence of pests and diseases and the nature of the weather warnings that can be given. The uppermost portion of the calendar gives information related to the weather abnormalities and precautionary measures to be taken. This part is divided into different sections according to dry spell length, high wind, heavy rainfall and cloudy weather.

These calendars are useful for crop planning, irrigation scheduling and plant protection measures. Information on normal weather conditions, water requirement of crops, *etc*. are of vital importance for effective crop planning and for maximising and stabilising food production in the country. The diagrams in the calendar help the weather forecaster to see, at a glance, what warnings are to be issued for a particular district during a given weather situation during a particular phase of a crop. The calendars are being subjected to periodical checking and revision.

WEATHER FORECASTING METHODS

The methods in use for weather forecasting can be divided into three broad categories, *viz.*, conventional or synoptic, statistical and numerical weather prediction. The method that can be utilised for forecasting depends on the length of forecast, the type of information desired and how much is known about the current state of atmosphere.

Synoptic methods

It is the traditional approach in weather prediction. This primary method continued to be in use until the late 1950s.The present state of the atmosphere is observed and surface and upper air synoptic charts are prepared. Synoptic means the observation of different weather elements at a specific time. The inferences on expected movement of weather systems are drawn by using previous and present charts. In order to have an average view of the changing pattern of weather, a modern meteorological centre prepares a series of synoptic charts every day. Such synoptic charts form the very basis of weather forecasts. From the careful study of weather charts over many years, certain empirical rules were formulated. These rules helped the forecaster in estimating the rate and direction of the movement of weather systems. The experience of the meteorologists is the knowledge gathered over years of observations and the background theory on weather. Given a background condition, a particular weather pattern, *e.g.*, a low pressure area, tends to evolve on the average in a certain way. This may be termed as mean or 'climatological' behaviour. Also atmospheric fluctuations have a certain amount of 'persistency'. In other words, once a low pressure is formed, it persists for a while. Forecasts made by meteorologists based on 'climatology' and 'persistence' and empirical knowledge about evolution of weather systems are known as synoptic forecasts. Before the advent of computers, this was the only method of weather forecasting. Up to 12 hours or 24 hours in advance, the meteorologists can actually make very good forecasts based on this technique. But beyond that this type of forecast is not useful. In addition to synoptic charts, satellite pictures also supply considerable information on cloud spread to the forecaster in issuing short range forecasts. Short range forecasts are issued by using the past analogous situations on the presumption that a current situation will evolve on the lines of the past analogous one.

Statistical methods

The statistical methods are used mostly in long range and seasonal climate forecasts. These methods use the past records of weather data on the assumption that future will be a repetition of the past weather. The main purpose of studying the past weather data is to find out those aspects of the weather that are good indicators of the future events. After establishing these relationships, correct data can be safely used to predict the future conditions. Only overall weather can be predicted in this way. It is particularly of use in projecting only one aspect of the weather at a time.

Statistical techniques such as multiple regression and ARIMA (auto regressive integrated moving average) models are used for predicting Indian monsoon rainfall based on 16 global-land-ocean-atmospheric variables. A new 10 parameter Power Regression model is being used since 2003 for estimating monsoon rainfall at the national level. IMD advances the long range forecast by 40 days and generates a quantitative forecast update by mid-July. This information is very much useful for contingency planning. Forecasts can also be attempted in the absence of any data about current weather, provided that long term information is available.

Numerical weather prediction

Until the 1950s, weather analysis and forecasting depended entirely on the experience of the meteorologists and their interpretation of current and recent past conditions of the atmosphere. Numerical weather prediction (NWP) models have largely replaced the subjective and statistical prediction methods in the last few decades and these models are the preeminent tool of modern weather forecasting. NWP method is based on computer programmes that attempt to mimic the actual behaviour of the atmosphere. The first successful numerical forecast was produced by Jule Gregory Charney and his associates in April, 1950. The state-of-the-art numerical models exhibit significant skill for forecast intervals as long as about a week.

The motion of the air is governed by a set of physical laws. These laws determine the evolution of the atmosphere from a given state to some future state. For most, but not all, meteorological problems the relevant laws of physics can be expressed in terms of seven equations which govern the behaviour of seven variables. The NWP requires enormous computations with about 1.3 million points for these variables. If the initial states of these variables are known and the equations are solved, the future states could be predicted. These equations are highly nonlinear coupled with partial differential equations.

Table 13.3 Variables and equations that define the state of atmosphere

Variables	Equations	Applications
Pressure (P)	Gas law or equation of state	Relates temperature, pressure and density
Temperature (T)	First law of thermo-dynamics	Relates temperature changes to external heating or cooling due to radiative balance and latent heat released during cloud formation
Density (ρ)	Continuity equation for air	Expresses conservation of mass of air
Water vapour (q)	Continuity equation for water vapour	Expresses rate of change of water vapour in terms of source (evaporation) and sink (precipitation) and diffusion due to turbulent eddies
West to east component of wind (u)	Newton's second law (force = mass × acceleration)	Applied to west-east direction; the pressure gradient force, the Coriolis force, gravitational force and frictional force
South to north component of wind (v)	Newton's second law	Applied to south-north direction; the pressure gradient force, the Coriolis force, gravitational force and frictional force
Vertical component of wind (w)	Newton's second law	Applied to vertical direction; the pressure gradient force, the Coriolis force, gravitational force and frictional force

Weather agencies around the world develop their own models and typically maintain a suite of models which are continually updated. Although there are large differences among models, the general procedure is the same for all numerical models and each has three phases, *i.e.*, analysis, prediction and post processing. In the analysis phase, observations are used to supply values corresponding to the starting state of the atmosphere for all the variables carried in the models. Analysis involves integration of the latest available observations, past values and outputs from other models to generate uniform initial values throughout the atmosphere over land and ocean. In the prediction phase, the values delivered by the analysis phase are used in the governing equations to obtain new values a few minutes into the future. The process is then repeated, using the output from the first step as input for the next set of calculations. This procedure is performed over and over as many times as necessary to reach the end of forecast period. Governing equations cannot be solved directly and are broken down into simple operations that computers can perform. The forecast conditions at regular intervals (*e.g.*, every 12 hours) are represented on a grid for mapping. In the post processing phase, such maps for each period from different models are studied to recognise the pattern and the final forecast is to be made.

The NWP technique is used mostly in medium range weather forecasting. This method has a practical limit of 3 to 5 days in the tropical region and about 10 days in mid-latitudes. The NCMRWF, New Delhi gives medium range weather forecasting by using supercomputers. The NWP techniques are also used for short range forecasting of meteorological parameters.

APPLICATION OF AGROMETEOROLOGICAL FORECASTS

Meteorological forecasts for agricultural production can be provided for making strategic, tactical, and operational decisions. Lead time and number of days the forecast represents are the variables that determine whether a forecast is used for strategic, tactical, or operational decisions. Climatological forecasts are used in making strategic decisions. Both meteorological and climatological forecasts are used in making tactical decisions, and only meteorological forecasts are used in making operational decisions (Hollinger, 2009).

Strategic decision

Strategic agricultural climatological forecasts provide the data and information for planning agricultural practices that change on a time scale of months to years. These include selection of crop or livestock species, crop rotation, size of farm, size and type of equipments, tillage practices, planting date and spacing, cultivar selection, livestock stocking of pasture or range, design of livestock facilities, and investment decisions such as purchase of land, irrigation systems, and storage facilities. Crop forecasts for strategic decisions require application of crop models to either existing climate data or forecast climate data. In practice, either historical daily weather data or weather data from 'weather generators' are used with the assumption that the historical or generated data represent the future weather mean and variance. 'Weather generators' are computer programmes that create sequences of daily weather data fitting to the known distribution of each weather variable.

Tactical decision

Tactical agricultural meteorological forecasts provide meteorological data with lead times of months. These forecasts can be used to plan final changes in crop cultivar selection, timing of planting and harvest, planting density and spacing, timing of pest and disease control measures, intercultural operations, irrigation schedules, nutrient management, pre-season crop marketing decisions, and purchase or disposal of livestock.

Operational decision

The transition of forecasts from tactical to operational decisions is, in some cases, difficult to define. This is because forecasts for both tactical and some operational agricultural production decisions include current weather observations, with forecasts of the important variables into the future. A major difference is that the decisions being made in an operational framework are generally days into the future, rather than weeks to months for tactical decisions. Such decisions can include timing of planting, cultivation, and harvest; irrigation scheduling; fertiliser and pesticide application; and livestock movement between pastures, or to shelter.

WEATHER SERVICES TO FARMERS

Accurate forecasts are vital for the farming because weather deeply affects agricultural operations like risk assessment, crop planning, sowing, pest management, manuring, harvesting and storage. Accurate long term forecasts prompt efficient crop planning. The medium range forecasts focussing on major weather events can allow a farmer to more precisely plan critical decisions such as sowing, pesticide and fertiliser application, and harvesting. To meet the farmers' need in real-time the India Meteorological Department (IMD) has started Integrated Agromet Advisory Service (IAAS) in the country involving all the concerned organisations, *viz.*, Indian Council of Agricultural Research (ICAR), Ministry of Agriculture (both Centre and State), State Agricultural Universities (SAUs) and other agencies from April 2007.

National agromet advisory bulletin

National Agromet Advisory Service Centre (NAASC) has been set up in Agrimet Division of IMD at Pune to act as the nodal centre and headquarters of state Agromet Advisory Services (AAS) units of IMD. Since July 2004 this centre has been preparing All India Agromet Advisory Service bulletins on Thursday for issuing the same to the Ministry of Agriculture, Ministry of Earth Sciences and to the members of Crop Weather Watch Group (CWWG), who meet every Friday at Krishi Bhavan, Ministry of Agriculture, Govt. of India, New Delhi. Based on these inputs, important decisions on short time crop planning in various parts of the country are taken in this meeting. Besides, these bulletins are also uploaded on real time basis in the website (http://imdagrimet.gov.in) of the Agricultural Meteorological Division, IMD, Pune. Monthly and seasonal reports on Agromet Advisory Services are also prepared.

State level composite bulletin

The state level composite AAS bulletins are issued twice a week (Tuesday and Friday) from 23 AAS units of IMD located at different state capitals on operational basis in consultation with the Director of Agriculture of the respective states to prepare state level agricultural planning and management. The bulletins are disseminated in English as well as in local languages through radio, television, newspaper, internet, *etc.* The state level bulletin is a composite of district bulletins helping to identify the distressed district of the state as well as plan the supply of appropriate farm inputs such as seeds, irrigation water, fertiliser, pesticides, *etc*.

District agromet advisory service bulletin

IMD has started issuing quantitative district level weather forecast up to 5 days from 1st June, 2008. The products comprise of quantitative forecasts for 7 weather parameters, *viz*., rainfall, maximum and minimum temperatures, wind speed and direction, relative humidity and cloudiness. In addition, weekly cumulative rainfall forecast is also provided. IMD, New Delhi generates these products based on a Multi Model Ensemble technique using forecast products available from models developed in India and other countries. The products are communicated to 130 Agromet Field Units (AMFUs) located at SAUs and ICAR institutes covering all agroclimatic zones of the country. On receipt of the forecasts, AMFUs prepare the medium range weather forecast based agrometeorological advisories in vernacular language in consultation with a panel of experts in various disciplines of agriculture. These agroadvisories which are crop specific, weather event specific and farm operation specific are disseminated to the farmers through all possible mass media like newspaper, radio, television and also through personal contacts by extension workers. The advisories are kept as simple as possible both in terms of the language and terminology, in view of literacy level of the farming community. In addition to the farmers, these bulletins are also provided to authorities of concerned departments like those of Agriculture, Horticulture, Irrigation, Soil Conservation, Animal Husbandry, *etc*. to enable them to take necessary measures for effective utilisation of the advisories.

CHALLENGES AND COMPLEXITIES OF WEATHER FORECASTING

Weather observation techniques have improved and there have been technological advancements in predicting the weather in recent times. On average, a five-day weather forecast of today is as reliable as a two day weather forecast 20 years ago. Despite this major scientific and technical progress, many challenges remain unsolved regarding long-term weather predictability. Weather forecasts for today or tomorrow are likely to be more dependable than predictions about the weather about two weeks from now. Sometimes it is also felt that weather forecast accuracy falls significantly beyond about 10 days. The accuracy of individual weather forecasts varies significantly.

Weather forecasting is a complex and often challenging skill that involves observing and processing vast amounts of data. Forecasting is a three step process; observing, forecasting, and communicating. Short range forecasting, looking to a few hours into the future, typically depends on closely observing how weather systems are currently evolving and tracking, and projecting their movement into the future based on what is understood about dynamics of the atmosphere. Forecasting beyond a day rely more on numerical weather modelling. Numerical forecast models compile data from surface observations, weather balloons and satellite imagery to create a computer generated simulation of the weather into the future. The model simulations use dynamic equations that express how the atmosphere will respond to changes in temperature, pressure and humidity over time. However, the earth's atmosphere is a complicated system that is affected by many factors and can react in different ways.

Once a forecast is created with reasonable confidence, the next role of the forecaster is to deliver that forecast in a way that the users can understand and appropriately respond to. The timeliness of issuing the forecast to provide adequate response to users is also one of the concerns. The medium range weather forecasting based agroadvisories should be location specific so that suitable actions can be taken timely for altering or prompting the agricultural operations. The most important challenges that the forecasters face today are to improve the skill of forecasting through better understanding of atmospheric phenomena and thus improving the relevant processes, and to increase the lead time of the agrometeorological forecast from the present 3 to 4 days to at least one week. This will be useful in the field of agriculture and animal husbandry in a big way.

REMOTE SENSING

The word 'remote sensing' was coined by Evelyn L. Pruitt, an American Geographer in 1960. Remote sensing is defined as collection and interpretation of information about a target without being in physical contact with it. Remote sensing is the science and art of obtaining information about an object, area or phenomenon through the analysis of data acquired by a device that is not in contact with the object, area or phenomenon under investigation (Lillesand and Keifer, 1994).

The information needs a physical carrier to travel from the objects/events to the sensors through an intervening medium. Any force field such as gravity, magnetic or electromagnetic could be used for remote sensing. However, the electromagnetic radiation is conventionally used as an information carrier in remote sensing. Remotely sensed data may be gathered by different devices, including sensors, film cameras, digital cameras and video recorders. Instruments capable of measuring electromagnetic radiation are called sensors.

Remote sensing is nothing but opposite of astronomy. In astronomy, man looks from the earth to the sky to understand and gain more knowledge about the celestial objects. But remote sensing refers to going upward in the sky to look downwards to understand and gain more knowledge about terrestrial objects. In remote sensing, the aerial sensor technologies are used to detect and classify objects on earth by means of electromagnetic radiation.

Satellite remote sensing technology is used to gather information on various aspects and for various purposes. The information thus obtained is superior to that obtained through simple visual techniques with respect to extent of coverage, permanency of data recorded, spectral and spatial resolution, speed and consistency in data interpretation, cost effectiveness, reliability, *etc.*

Advantages of remote sensing

1. Relatively cheap and rapid method of acquiring up-to-date information over a large geographical area.
2. It is the only practical way to obtain data from inaccessible regions, *e.g.*, Antarctica, Amazonia.
3. At small scales, regional phenomena which are invisible from the ground are clearly visible.
4. It is cheap and rapid method of constructing base maps in the absence of detailed land surveys.
5. It obtains imagery of any area over a continuous period of time through which any anthropogenic or natural changes in the landscape can be analysed.
6. Aerial and satellite data are permanent and reliable record of ground features present at the time of sensing by multispectral scanners.
7. Remote sensing technique offers greater fineness of image with much accuracy relative to traditional methods.
8. With the use of digital image processing system, data can be consistently and speedily analysed.

Limitations of remote sensing

1. Data collected through remote sensing are not direct samples of the phenomenon, so must be calibrated against reality.
2. Data must be corrected geometrically and georeferenced in order to be useful as maps, not only as pictures. This can be easy or complicated.
3. Distinct phenomena can be confused if they look as the same to the sensor, leading to classification error; *e.g.*, artificial and natural grasses in green light are not distinguished. However, infrared light can easily distinguish them.
4. Phenomena which were not meant to be measured for an application can interfere with the image and must be accounted for.

5. The interpretation of imagery requires a certain level of skill.
6. Data from multiple sources may create confusion and need cross verification with ground survey data.
7. Remotely sensed data are too complicated to use and sometimes may be misclassified or confused.
8. Remotely sensed data are not readily available and for low volume of data, use of remote sensing may be too expensive.
9. Distortions may occur in an image due to the relative motion of sensor and source.

Applications of remote sensing in agriculture

Remote sensing techniques are used for survey of natural resources like agriculture, forestry, water resources, soil mapping, land classification, *etc.* It can also be applied in tropical cyclone storm surveillance, monsoon studies and weather forecasting.

Crop mapping: The satellites have ability to image individual fields and regions on a frequent revisit cycle. The field-based information including crop identification, crop area determination and crop conditions can be obtained by remotely sensed data. Satellite data are employed in precision agriculture to manage and monitor farming practices at different levels. The data can be used for farm optimisation and spatially-enabled management of technical operations. The images can help determine the location and extent of crop stress and then can be used to develop and implement a spot treatment plan that optimises the use of agricultural chemicals. The major agricultural applications of remote sensing are crop type classification, crop condition assessment, crop monitoring and damage assessment, and crop yield estimation. By using stratified sampling methodology crop acreage estimation can be done to the high level precision. Remote sensing technology can be used to prepare maps of crop type and delineating their extent. It can generate a systematic and repetitive coverage of a large area and provide information about the health of the vegetation. Crop yields are estimated by analysing satellite based vegetation indices which are transformations of reflectance in the near infrared portions of electromagnetic spectrum. The crop conditions affected by the factors like deficiency of nutrients, acidity or salinity of soil, adverse weather conditions, *etc.* can be detected by remote sensing. Infestation by insect pests and diseases can also be identified by remote sensing. The data of crop is used to prepare an inventory of what was grown in certain areas and when. This information serves to predict crop yield, collecting crop production statistics, facilitating crop rotation records, mapping soil productivity, identification of factors influencing crop stress, assessment of crop damage and monitoring farming activity.

Water resource mapping: The damage of crops due to floods, cyclones and water logging can be detected and managed effectively by using techniques like multi-temporal remote sensing. The surface and sub surface water availability for irrigation and the amount of moisture stored in the upper few centimetres of soil can also be estimated to a greater accuracy. Also, it facilitates in delineation of watershed areas. Assessing the drought realistically and ways to manage the adverse effects is possible through remote sensing.

Soil mapping: The disturbance of soil by land use impacts on the quality of the environment. Salinity, soil acidification and erosion are some of the problems. The remote sensing technology is highly useful in identifying, mapping and delineating degraded lands. Soil layers that rise to the surface during erosion have different colour, tone and structure than non eroded soils, thus the eroded parts of soil can be easily identified on the images. Remote sensing provides information to identify the important indicators of desertification. Based on this, action can be taken by the planners at different levels.

Forest mapping: Deforestation is one of the most important global problems. There are many implications of it. The loss of forests increases soil erosion, river siltation and deposition, affecting the environment. The forest resources can be monitored by remote sensing data. Identification of forest cover typing (canopy texture, leaf density) and tree species is one of the basic applications of remote sensing. Forest cover typing can consist of reconnaissance mapping over a large area, while species inventories are highly detailed measurements of stand contents and characteristics (tree type, height, density). Various forest types can be identified and delineated by using remote sensing data that would be difficult and time consuming using traditional ground surveys. Forest fire can also be detected by using remote sensing and corrective measures can be taken timely.

Limitations of remote sensing application in agriculture

There are some operational limitations in applying remote sensing technology in agriculture in India. Cloud cover during monsoon season, the major crop growing season in India, restricts data gathering by remote sensing. Classification of images sometimes becomes difficult due to poor growth of the crops since large area comes under rainfed farming with poor agronomic management practices. Small landholdings with smaller cropped fields, mixed cropping and different cultural practices to the crops also affect quality of remotely sensed data. Appearance of some crops (like wheat and barley) makes confusion for identification. These limitations must be kept in mind while applying remote sensing data in agriculture.

14

Climate Change

Many aspects of the global climate are changing rapidly. Evidence for changes in the climate system abounds, from the top of the atmosphere to the depths of the oceans. The climate system is the highly complex system consisting of five major components; the atmosphere, the hydrosphere, the cryosphere, the land surface and the biosphere, and the interactions among them. The climate system evolves in time under the influence of its own internal dynamics and because of external forcings such as volcanic eruptions, solar variations and human-induced forcings such as the changing composition of the atmosphere and land-use.

The term 'climate variability' is often used to denote deviations of climatic statistics over a given period of time (*e.g.*, a month, season or year) from the long-term statistics relating to the corresponding calendar period. In this sense, climate variability is measured by those deviations, which are usually termed anomalies.

'Climate change' refers to a statistically significant variation in either the mean state of the climate or in its variability, persisting for an extended period (typically decades or longer). Climate change may be due to natural internal processes or external forcings, or due to persistent anthropogenic changes in the composition of the atmosphere or in land use. The United Nation's Framework Convention on Climate Change (UNFCCC) defines 'climate change' as 'a change of climate behaviour which is attributed directly or indirectly to human activity that alters the composition of the global atmosphere and which is in addition to natural climate variability observed over comparable time periods'. The UNFCCC thus makes a distinction between 'climate change' attributable to human activities altering the atmospheric composition, and 'climate variability' attributable to natural causes. Climate change is a long-term shift in weather conditions measured by changes in temperature, precipitation, wind, snow cover, and other indicators. It can involve both changes in average conditions and changes in variability, including, for example, changes in extreme conditions.

A key difference between climate variability and change is in persistence of 'anomalous' conditions. In other words, events that are used to be rare occur more frequently, or *vice versa.* Occasionally, an event or sequence of events occurs that has never been recorded before, such as the exceptional tsunami in the Indian Ocean in 2004 or hurricane season in the Atlantic in 2005. Yet even that could be a part of natural climate variability. If such a season does not recur within the next 30 years, by looking back it could be called as an exceptional year, but not a 'climate change'. Only a persistent series of unusual events taken in the context of regional climate parameters can suggest that a potential change in climate has occurred.

CAUSES OF CLIMATE VARIABILITY AND CHANGE

Climate change is a movement in the climate system because of internal changes within the climate system or in the interaction of its components, or because of changes in external forcing either by natural factors or anthropogenic activities (IPCC, 1996). Natural variability is a characteristic of the global climate and occurs on both long and short time scales. s

Some external influences, such as changes in solar radiation and volcanism, occur naturally and contribute to the total natural variability of the climate system. Other external changes, such as the change in composition of the atmosphere that began with the industrial revolution, are the result of human activity. The influence of solar irradiance changes and volcanic eruptions has been very small relative to the influence of human emissions of greenhouse gases since the beginning of the industrial revolution. The build-up of greenhouse gases in the atmosphere is the primary cause for concern about climate change now and into the immediate future.

The causes of climate change can broadly be divided into two categories, natural and human causes. The natural variability and the climate fluctuations of the climate system have always been part of the earth's history; however, there have been changes in concentrations of greenhouse gases in the atmosphere growing at an unprecedented rate and magnitude.

Natural causes

The earth's climate can be affected by natural factors, such as volcanic eruptions, ocean current, the earth's orbital changes and solar variations.

Volcanic eruptions: When a volcano erupts it throws out large volumes of sulphur dioxide, water vapour, dust and ash into the atmosphere. Large volumes of gases and ash can influence climatic patterns for years by increasing planetary reflectivity causing atmospheric cooling. Tiny particles called aerosols are produced by volcanoes. Because they reflect solar energy back into space they have a cooling effect on the world. Volcanic eruptions are episodic and have relatively short-term effects on climate.

Ocean current: The oceans are a major component of the climate system. Ocean currents move vast amounts of heat across the planet. Winds push horizontally against the sea surface and drive ocean current patterns. Interactions between the ocean and atmosphere can also produce phenomena such as *El Nino* which occurs every 2 to 6 years. The oceans play an important role in determining the atmospheric concentration of CO_2. Changes in ocean circulation may affect the climate through the movement of CO_2 into or out of the atmosphere.

Earth's orbital changes: The earth makes one full orbit around the sun each year. It is tilted at an angle of 23.5° to the perpendicular plane of its orbital path. Changes in the tilt of the earth can lead to small but climatically important changes in the strength of the seasons; more tilt means warmer summers and colder winters; less tilt means cooler summers and milder winters. Slow changes in the earth's orbit lead to small but climatically important changes in the strength of the seasons over tens of thousands of years.

Solar variations: The sun is the source of energy for the earth's climate system. Although the sun's energy output appears constant from an everyday point of view, small changes over an extended period of time can lead to climate changes. As the sun is the fundamental source of energy that is instrumental in our climate system it would be reasonable to assume that changes in the sun's energy output would cause the climate to change. Current global warming however cannot be explained solely by solar variations.

Human causes

Climate change can also be caused by human activities, such as the burning of fossil fuels and the conversion of land for forestry and agriculture. Since the beginning of the industrial revolution in the 19th century, these human influences on the climate system have increased substantially. Carbon dioxide is the most important greenhouse gas in the atmosphere. Changes in land use pattern, deforestation, land clearing, agriculture, and other activities have all led to a rise in the emission of carbon dioxide. Methane is another important greenhouse gas in the atmosphere. It is released from animals such as dairy cows, goats, pigs, buffaloes, *etc*. Methane is also emitted during the process of oil drilling, coal mining, leaking gas pipelines, landfills and waste dumps. It is this human-induced enhancement of the greenhouse effect that is of concern because ongoing emissions of greenhouse gases have the potential to warm the planet to levels that have never been experienced in the history of human civilization. This change in the climate parameter, also known as global warming, has occurred faster than any other climatic change recorded by humans. Such climate change could have far-reaching and/or unpredictable environmental, social, and economic consequences.

Greenhouse gases

The earth has a natural greenhouse effect where certain gases, known as greenhouse gases, in the atmosphere allow the sunlight to enter and absorb the heat radiation. Because these gases absorb the heat, they keep the average surface temperature on earth around 14° C. Without the natural greenhouse effect, the earth's average surface temperature would be around –19° C. Since the industrial revolution, human activity has increased the amount of greenhouse gases in the atmosphere. The increased amount of gases which absorb heat, has directly lead to retention of more heat in the atmosphere and thus an increase in global average surface temperatures. This change in temperature is known as global warming. The increase in temperature is also leading to other effects on the climate system. Together these effects are known as anthropogenic climate change. The three main causes of the increase in greenhouse gases observed over the past 250 years have been fossil fuels, land use and agriculture (IPCC, 2006).

The principal greenhouse gases (GHGs) include carbon dioxide, methane, nitrous oxide and chlorofluorocarbons (CFCs). Carbon dioxide constitutes only a small part of the atmosphere, but it is one of the most important GHGs. CO_2 is released naturally into the atmosphere through volcanic eruptions and animal respiration but it is also released through human activities such as deforestation and the burning of fossil fuels for energy. CO_2 also spends a long time in the atmosphere increasing its impact. Methane, the second most important GHG, is produced both naturally and through human activities. The most significant sources of methane come from the decomposition of organic matter under anaerobic conditions. Another large source is from the digestion of ruminants. Methane is a stronger GHG than CO_2 because it can absorb more heat; however it is much less abundant in the atmosphere. Nitrous oxide, a very powerful greenhouse gas is released during the production and use of organic fertilisers. It is also produced when burning fossil fuels. Chlorofluorocarbons are the manmade compounds which are mainly used in refrigerants and air conditioners.

Since the beginning of the 20th century industrial activity grew 40-fold, and the emissions of greenhouse gases grew 10-fold. The amount of CO_2 in the air increased from some 280 parts per million by volume (ppmv) at the beginning of the century to 389 ppmv at the end of 2010. It is predicted that the level will be 450 ppmv in 2050 resulting in increase in temperature by 1.8 to 3° C. The amount of CO_2 varies throughout the year as the result of the annual cycles of photosynthesis and oxidation. Similarly, methane rose from a pre-industrial atmospheric concentration of around 700 parts per billion by volume (ppbv) to about 1789 ppbv by 2007. The change in major green house gases since the industrial revolution is presented in the table 14.1.

Table 14.1 Change on major greenhouse gases since the industrial revolution

Gas	Pre-1750 tropospheric concentration	Recent tropospheric concentration (2012)	Increased radiative forcing (W/m^2)
Carbon dioxide (ppm)	280	395.4	1.88
Methane (ppb)	722	1893	0.49
Nitrous oxide (ppb)	270	326	0.17
Tropospheric ozone (ppb)	237	337	0.40

(*Source*: IPCC, 2013)

Aerosols in the atmosphere

Atmospheric aerosols are able to alter climate in two important ways; they scatter and absorb solar and infrared radiation and may change the microphysical and chemical properties of clouds and possibly their lifetime and extent. The scattering of solar radiation acts to cool the planet, while absorption of solar radiation by aerosols warms the air directly instead of allowing sunlight to be absorbed by the surface of the earth. Human activity contributes to the amount of aerosols in the atmosphere in several ways. Dust is often a bi-product of agricultural processes. Biomass burning and industrial processes produce a wide variety of aerosols depending on the type of raw materials and products. Exhaust emission from transports generates pollutants that are either aerosols from the outset, or are converted by chemical reactions in the atmosphere to form aerosols.

Land use change

Land-use changes (*e.g.*, cutting down forests to create farmland) have led to changes in the amount of sunlight reflected from the ground back into space (the surface albedo). The scale of these changes is estimated to be about one-fifth of the forcing on the global climate due to changes in emissions of greenhouse gases. The largest effect of deforestation is estimated to be at high latitudes where the albedo of snow-covered land, previously forested, has increased. This is because snow on trees reflects only about half of the sunlight falling on it, whereas snow-covered open ground reflects about two-thirds. Another contributing cause of climate change is when agriculture alters the earth's land cover, which can change its ability to absorb or reflect heat and light. Land use change such as deforestation and desertification, together with use of fossil fuels, are the major anthropogenic sources of carbon dioxide release into the atmosphere.

Deforestation

The rainforests form part of a delicate ecosystem that has taken millions of years to evolve. Rainforests every year help to absorb almost 20% of manmade CO_2 emissions; therefore, deforestation can be classed as a major contributor to the causes of climate change. Cutting down rainforests faster than they can be replaced has a devastating effect on the carbon emission cycle producing an extra 17% of greenhouse gases. Deforestation by means of cutting down and burning these tropical rainforests usually pave the way for agriculture and industry which often produce even more CO_2. Fossil fuels release carbon dioxide into the atmosphere contributing to global warming and climate change. Forest alleviates this change by converting carbon dioxide to carbohydrate during photosynthesis. The world's forests contain about 125% of the carbon found in the atmosphere. This carbon is stored in the form of wood and vegetation through 'carbon sequestration'. Trees possess about 20% carbon by weight and biomass of forest acts as a 'carbon sink'. The soil organic matter produced by the decomposition of dead plant material also acts as a carbon store.

CONTRIBUTION OF LIVESTOCK TO CLIMATE CHANGE

Ruminant animals, such as cattle, sheep, buffalo, and goats have special digestive systems that can convert otherwise unusable plant materials into nutritious food and fibre. This helpful digestive process, called enteric fermentation, however, produces methane (CH_4), a potent greenhouse gas that can contribute to global climate change. Enteric fermentation is a digestive process by which carbohydrates are broken down by microorganisms into simple molecules for absorption into the bloodstream of an animal. It is one of the factors in increased methane emissions. Enteric fermentation occurs when methane is produced in the rumen as microbial fermentation takes place.

Livestock production systems can also emit other greenhouse gases such as nitrous oxide (N_2O) and carbon dioxide (CO_2). In recent years the increasing use of intensive livestock production systems has become a source of solid, liquid and airborne emissions that can be both a nuisance and environmentally harmful. In spite of the low amount of CH_4 in the atmosphere relative to that of CO_2, its importance as a pollutant is considered to be 21 times greater than that of CO_2, while that of N_2O is 310 times greater (Hartung, 2003).

It is estimated that nearly 20% of CH_4 and 35% of N_2O come from livestock production. According to the European Environment Agency, nearly 50% of the overall amount of CH_4 released in Europe originates from agriculture and ruminant animals (EEA, 2001).

Table 14.2 Relative contribution(%) of various sources to the global emission of methane (CH_4) and nitrous oxide (N_2O)

Gas	Natural sources	Anthropogenic sources	
		Livestock production	Others
CH_4	30	20	50
N_2O	55	35	10

(*Source*: Monteny, 2003)

Manure of cattle is a much more important source of CH_4 emission than enteric fermentation. This problem is due to liquid manure storage in tanks or pits. The amount and surface area of the stored manure, and the ambient and core manure temperature are the major factors that determine the quantity of CH_4 gas emission (Hartung and Monteny, 2000).

Table 14.3 Methane emission from livestock production (kg/animal/year)

Species	Enteric fermentation	Manure
Dairy cows	100	345
Pigs	1	32
Poultry	0.1	2.4

(*Source*: Hartung, 2003)

Thus, the livestock sector which will be a sufferer of climate change is itself a large source of methane emissions. In India, although the emission rate per animal is much lower than the developed countries, due to vast livestock population the total annual methane emissions from Indian livestock ranged from 7.26 to 10.4 million tons/year. In India more than 90% of the total methane emission from enteric fermentation is being contributed by the large ruminants (cattle and buffalo) and rest from small ruminants and others.

Reducing impacts of livestock production on climate

Methane is generated mainly as a by-product of the fermentation of the digestible organic matter in ruminants; especially those with forage-based diets. The grain-based diets reduce the emission of CH_4. Thus, animal diet composition is an important influencing factor to determine the volume of CH_4 gas emission. To reduce the emissions of CH_4 and N_2O the following strategies should be considered (WMO, 2010).

1. Replacement of roughage in the cattle diet with concentrates

2. Development of low-emission production system facilities, including filters, scrubbers, covered manure pits and shallow manure application
3. Reduction in the concentration of animals in intensive production units to the extent possible, by using more pens and pastures
4. Use of feed additives to reduce CH_4 emissions; lower amounts of nitrogen in manure and urine can reduce N_2O emissions
5. Increase in feed digestibility and feed conversion efficiency

CLIMATE CHANGE SCENARIO IN INDIA

Climate change in India may pose additional stresses on ecological and socioeconomic systems that already face tremendous pressures from rapid urbanisation, industrialisation and economic development. India Meteorological Department (IMD) maintains a well distributed network of more than 500 stations in the country for more than a century. Meteorological data collected from these stations from 1901 to 2009 was analysed and the trend in change of climate parameters in India was published (Attri and Tyagi, 2010).

Temperature

The annual mean temperature for the country as a whole has risen by 0.56° C over a period of 1901-2009. It may be mentioned that annual mean temperature has been generally above normal (normal based on period, 1961-1990) since 1990. This warming is primarily due to rise in maximum temperature across the country, over larger parts of the data set. However, since 1990, minimum temperature is steadily rising and rate of its rise is slightly more than that of maximum temperature. Spatial pattern of trends in the mean annual temperature shows significant increasing trend over most parts of the country except over parts of Rajasthan, Gujarat and Bihar, where significant decreasing trends were observed. Season wise, maximum rise in mean temperature was observed during the post-monsoon season (0.77° C) followed by winter season (0.70° C), pre monsoon season (0.64° C) and monsoon season (0.33° C).

Precipitation

The country as a whole does not show any significant trend with respect to the all India annual and monsoon rainfall for the period 1901-2009. Similarly rainfall for the country as whole for the same period for individual monsoon months also does not show any significant trend. However, during the winter season, rainfall is decreasing in almost all the subdivisions except in Himachal Pradesh, Jharkhand, Nagaland, Manipur, Mizoram and Tripura. Rainfall is decreasing over most parts of the central India during the pre-monsoon season. However during the post-monsoon season, rainfall is increasing for most of the sub-divisions.

Extreme rainfall events

A large amount of the variability in rainfall is related to the occurrence of extreme rainfall events. Number of heavy rainfall events is increasing almost over the entire landmass. Also the frequency and intensity of extreme events defined as one-day maximum precipitation shows increasing trend everywhere except some northern parts of the country. Country's highest observed one day point rainfall (156.3 cm) and also world's highest 2-day point rainfall (249.3 cm) occurred in Cherrapunji in the year 1995 (IMD, 2006). The exceptionally heavy rainfall of 944 mm over Mumbai on 26th July, 2005 was very unprecedented in nature, which led to many more studies on frequency and variability of heavy rainfall events.

GENERAL EFFECTS OF CLIMATE CHANGE

Climate is changing naturally at its own pace since the beginning of evolution of the earth. But presently it has gained momentum due to inadvertent anthropogenic disturbances. Today, climate change is a global concern. The earth is facing one of the greatest environmental, economic and social threats through climate change. Stronger and more frequent typhoons, warmer nights, and longer and hotter days are some of the most common evidences of climate change. The general consequences of climate change are discussed below.

Ecological imbalance: Climate change can make summer months warmer or cold months cooler than usual. These changes in temperature can cause animals to migrate to more suitable places or force them to adapt with adverse effects on their physical conditions. Some may even die and become extinct because of their inability to adapt to extreme environmental conditions. Biodiversity is thus at more risk now than ever because of climate change.

Extreme climate variability: Climate variability refers to short term changes in climate such as longer dry or rainy season, intense heat during summer, more or scanty rains during rainy months.

Increasing temperature: Global warming is the biggest threat being faced by mankind today. It refers to the observed and projected increase in the earth's average temperature due to natural or anthropogenic causes. The measured average atmospheric temperature near the earth's surface rose by 0.74° C during the last 100 years, 1906–2005 (IPCC, 2007). Cold days, cold nights and frost have become less frequent and hot days, hot nights and heat waves more frequent. Diurnal temperature range has also decreased. It is estimated that average annual temperature will rise by 1.8 to 4° C within the next 100 years. Gases created through industrial and agricultural practices increase the heat-reflecting potential of the atmosphere, thereby raising the planet's average temperature.

Limited food supply: Climate change can alter the schedule and duration of cropping seasons. Farmers can become confused as to when they should plant crops, thereby affecting length of cropping season, time of harvest and food supply. Water shortage during dry months can also affect crop growth and overall food production.

Increased occurrence of pests and diseases: Climate change can affect the life cycle of pests by increasing their population at a different time. Farmers may be unaware of or caught unprepared for these changes. Diseases can also become prevalent based on the environmental conditions resulting from climate change.

More powerful typhoons: Global warming increases ocean temperature and rate of evaporation. As water vapour in the atmosphere increases, the typhoons carry more rains. Climate change, through global warming has increased the frequency and intensity of typhoons over the years.

Rise in sea levels: Increasing global temperature causes glaciers and polar icecaps to melt, thereby making sea levels rise. If sea levels continue to increase, more land could be submerged under water permanently. The IPCC estimates that sea levels rose 17 cm in 20th century and it may further increase by 18-59 cm by the year 2100.

Glacier retreat and disappearance: The continued retreat of glaciers will have a number of different quantitative impacts. In the areas that are heavily dependent on water runoff from glaciers that melt during the summer months, a continuation of the current retreat will eventually deplete the glacial ice and substantially reduce or eliminate runoff. A reduction in runoff will affect the ability to irrigate crops and will reduce summer stream flows necessary to keep dams and reservoirs replenished. Approximately 2.4 billion people live in the drainage basin of the Himalayan rivers. India, China, Pakistan, Afghanistan, Bangladesh, Nepal and Myanmar could experience floods followed by severe droughts in coming decades. In India alone, the Ganges provides water for drinking and farming for more than 500 million people.

EFFECT OF CLIMATE CHANGE ON AGRICULTURE

Agriculture is the largest human activity in the world which depends on climatic parameters. Climate change and agriculture are interrelated processes, both of which take place on a global scale. Despite technological advances, such as improved crop varieties, genetically modified organisms, and irrigation systems, weather is still a key factor in agricultural productivity, as well as soil properties and natural communities. The effect of climate change on agriculture is related to variabilities in local climates rather than in global climate patterns. The growth, development and reproductive processes in both plants and animals are also temperature dependent. All these physiological processes show their maximum activity, hence good productivity, only at their optimum temperature. A genotype gives good yield only under suitable climatic conditions and other management practices. For each plant variety, there is an optimal temperature for vegetative growth, with growth dropping off as temperature increases or decreases. Similarly, there is a range of temperatures at which a plant will produce seed. Outside of this range, the plant will not reproduce. More than 50% differences in yield are due to climatic variations. Thus climate change has a profound effect on agricultural practices.

Altering cropping seasons, increasing incidence of pests and diseases, rationing water supply, loss of agricultural biodiversity and limiting food supply are all evidences of climate change. These factors force farmers to adapt using meagre resources. A significant effect of global climate change is the altering of rainfall patterns, which certainly affects agriculture. Extended drought can cause the failure of small and marginal farms with resultant economic, political and social disruption.

Impact of climate change on crop

Changes in temperature, amount of carbon dioxide (CO_2), and the frequency and intensity of extreme weather could have significant impacts on crop yields. Many weeds, pests and fungi thrive under warmer temperatures, wetter climates, and increased CO_2 levels.

Warmer temperatures may make many crops grow more quickly, but warmer temperatures could also reduce yields. Crops tend to grow faster in warmer conditions. However, for some crops, faster growth reduces the length of time available for seeds to grow and mature. In the case of an annual crop, the duration between sowing and harvesting will shorten. The shortening of such a cycle could have an adverse effect on productivity because senescence would occur sooner. Thus, yield potential of the crops may not be fully exploited. In some areas, warming may benefit the types of crops that are typically planted there. However, if warming exceeds a crop's optimum temperature, yields can decline. Increase in temperature also increases crop respiration rate and alters partitioning of photosynthates to economic products.

Higher CO_2 levels can increase yields. The yields for some crops, like wheat and soybeans, could increase by 30% or more under a doubling of CO_2 concentrations. However, some factors may counteract these potential increases in yield. For example, if temperature exceeds a crop's optimal level or if sufficient water and nutrients are not available, yield increases may be reduced or reversed.

More extreme temperature and precipitation can prevent crops from growing. Extreme events, especially floods and droughts, can harm crops and reduce yields. Dealing with drought could become a challenge in areas where summer temperatures are projected to increase and precipitation is projected to decrease. As water supplies are reduced, it may be more difficult to meet water demands. Extreme climate conditions may affect quality of fruits, vegetables, tea, coffee, aromatic and medicinal plants.

Decrease in stratospheric ozone may increase biologically dangerous ultraviolet radiation on earth's surface. Ultra violet radiation may damage nucleic acids creating long term effects. Excess ultraviolet radiation can affect plant physiology either directly or indirectly through changed pollinator behaviour and cause massive amounts of mutations. Tropospheric ozone causes foliar injury. This may reduce productivity of many crops.

More detailed analysis of rice yields by the International Rice Research Institute forecasts 20% reduction in yields over the region per degree (in Celsius) rise in temperature. Rice becomes sterile if exposed to temperatures above 35° C for more than one hour during flowering and consequently produces no grain.

Impact of climate change on soil health

The warmer atmospheric temperature observed over the past decades is expected to lead to a more vigorous hydrological cycle, including more extreme rainfall events. Erosion and soil degradation is more likely to occur, creating sedimentation in streams and reservoirs. Soil fertility would also be affected by global warming. The increase in the temperature would induce a greater rate in the N mineralisation but there will be more volatilisation and denitrification loss. High soil temperature reduces soil organic matter both in quantity and quality. This may also decrease microbial population reducing soil productivity and nutrient cycling. Excess rainfall causes waterlogging, soil salinity problems and salt water ingression in coastal areas.

Impact of climate change on water resources

Climate change will modify rainfall, evaporation, runoff and soil moisture storage. In some areas, more frequent heavy downpours may cause localised flooding. Increased evaporation from the soil and accelerated transpiration in the plants will cause moisture stress. Thus, crops will need frequent irrigation. The water resources will come under increasing pressure and could become seasonally scarce as a result of regional changes in water supply due to changes in precipitation or snow and glacier melt. Climate change affects the availability of fresh water for irrigation. The over draft and non-replenishment of groundwater will lower the groundwater level permanently.

Impact of climate change on pest

Global warming would cause an increase in rainfall in some areas, which would lead to an increase of atmospheric humidity and the duration of the wet seasons. Combined with higher temperature, these could favour the development of fungal diseases. Similarly, because of higher temperature and humidity, there could be an increased pressure from insects and disease vectors. An increase in temperature may accelerate the growth and multiplication of insect pests. The polycyclic diseases become epidemic with climate change. The geographical range of insects is expanded with increased risk of infestation.

Impact of climate change on livestock

Climate change, particularly global warming has complex impacts on domestic animal production system affecting feed supply, challenging thermoregulatory mechanism resulting thermal stress, emerging new diseases due to change in epidemiology of diseases and causing many other indirect impacts. So, climate change will have far-reaching consequences for dairy and meat production, especially in vulnerable parts of the world where it is vital for nutrition and livelihoods. Due to poor availability of good quality feed and fodders in India, animals are generally maintained on poor quality grasses available in the pastures or are stall-fed, mainly on crop residues. India is already deficit in feed and fodder. These shortages would be further aggravated by the adverse effects of global warming on fodder crop production. There will be less fodder production due to increased water scarcity. Rise in temperature causes more lignifications of plant tissues reducing its digestibility. Global warming increases the requirements for water, shelter and energy to meet the projected milk demands.

Drought may threaten pasture and feed supplies. Drought also reduces the amount of quality forage available to grazing livestock. Some areas could experience longer, more intense droughts, resulting from higher summer temperature and reduced precipitation. For animals that rely on grain, changes in crop production due to drought could also become a problem.

Heat waves, which are projected to increase under climate change, could directly threaten livestock. Heat stress affects animals both directly and indirectly. Heat stress is one of the greatest climatic challenges faced by the domestic animals and global climate warming may further aggravate the condition and even provoke thermal stress condition. Dairy cattle are particularly more susceptible to increased ambient temperature than other ruminants because of their high metabolic rate and poor water retention mechanism in kidney and gastrointestinal tract. Over time, heat stress can increase vulnerability to disease, reduce fertility, and reduce milk production.

Climate change may increase the prevalence of parasites and diseases that affect livestock. In areas with increased rainfall, moisture-reliant pathogens could thrive. Vector borne diseases may have more impacts on livestock. Increases in carbon dioxide may increase the productivity of pastures and plants on which livestock feed, but may also decrease their quality.

Impact of climate change on fisheries

Fishery enterprises already face multiple stresses, including overfishing and water pollution. Climate change may worsen these stresses. In particular, temperature changes could lead to significant impacts. Increased water temperature affects fish breeding, migration and harvest. Many marine species have certain temperature ranges at which they can survive. Increased temperature and tropical cyclonic conditions affect capture, production and marketing cost of marine fish. Many aquatic species migrate towards colder areas of streams, lakes or ocean to avoid warm water. However, moving into new areas may put these species into competition with other species over food and other resources. Changes in temperature and seasons could affect the timing of reproduction and migration. Many steps within an aquatic animal's lifecycle are controlled by temperature and the changing of the seasons. Some diseases that affect aquatic life may become more prevalent in warm water.

In addition to warming, the world's oceans are gradually becoming more acidic due to increases in atmospheric carbon dioxide. Increasing acidity could harm shellfish by weakening their shells, which are created from calcium and are vulnerable to increasing acidity. Acidification may also threaten the structures of sensitive ecosystems upon which some fish and shellfish rely. The effects of climate change on agriculture by 2050 as predicted by Ministry of Agriculture, Fisheries and Food, Govt. of United Kingdom are presented in table 14.4.

Table 14.4 Predicted effects of climate change on agriculture by 2050

Climatic element	Expected changes by 2050	Confidence in prediction	Effects on agriculture
CO_2	Increase from 360 ppm to 450–600 ppm	Very high	Good for crops; increased photosynthesis; reduced water use
Sea level rise	Rise by 10 -15 cm	Very high	Loss of land, coastal erosion, flooding, salinisation of groundwater
Temperature	Rise by 1-2° C; winters warming more than summers; increased frequency of heat waves	High	Faster, shorter, earlier growing seasons, range moving north and to higher altitudes, heat stress risk, increased evapotranspiration
Precipitation	Seasonal changes by ± 10%	Low	Impacts on drought risk, soil workability, water logging, irrigation supply, transpiration
Storminess	Increased wind speeds, more intense rainfall events	Very low	Lodging, soil erosion, reduced infiltration of rainfall
Variability	Increases across most climatic variables; predictions uncertain	Very low	Changing risk of damaging events (heat waves, frost, droughts, floods) which effect crops and timing of farm operations

(*Source*: MAFF, 2000)

MEASURES ADDRESSING THE EFFECT OF CLIMATE CHANGE ON AGRICULTURE

The adverse effects of climate change on agriculture have become a major course of concern in recent years. As agricultural production increases, resource constraints, particularly water, will become tighter. Agriculture globally accounts for about 70% of the world's freshwater withdrawals. Climate change is expected to alter the seasonal timing of rainfall and snow pack melt resulting in a higher incidence and severity of floods and droughts. The length of the growing season and the type of crop grown are both affected by changes in temperature. Both rainfed and irrigated agriculture need to be managed more sustainably to reduce production risks. Thus, to deal with effects of climate change, both adaptive measures and mitigation measures are to be taken.

Adaptation strategies

Adaptation may be defined as adjustment in natural or human systems in response to actual or expected climatic stimuli or their effects, which moderates harm or exploit beneficial opportunities. Adaptation as a response to change must be appropriate to specific hazards or threats in a given period of time; in the same way, an effective adaptation to a real or perceived change in local climate could, over time, become inappropriate as circumstances changes.

Adaptation in agriculture is the norm rather than the exception. Adaptation strategies will vary with agricultural systems, location, and scenarios of climate change considered. There are several adaptation measures that the agricultural sector can undertake to cope with future climate change. The greater uncertainty from climate change can be best addressed through contingent planning. Adaptation through such measures as crop and livestock insurance, social safety nets, and research on and dissemination of flood, heat and drought resistant crops, including conservation of traditional plant varieties with those characteristics can be facilitated.

Adaptive approaches to minimise the adverse effects of climate change on agriculture include initiating reforestation and afforestation activities, developing water harvesting techniques, improving irrigation efficiency, conserving soil moisture through appropriate tillage methods. Another adaptation strategy taken by farmers is crop diversification and changing planting dates that has proven to be one of the most popular farm level responses to climate variability and change. Improved livestock management practices like providing better housing and shade, change to heat tolerant breeds, change in stocking rate, altered grazing and rotation of pasture, *etc.* may prove to be beneficial. Early warning systems and protection measures for natural disasters like droughts, floods, tropical cyclones, *etc.* should be developed.

Mitigation technologies

The Intergovernmental Panel on Climate Change (IPCC, 2007) defines mitigation as the technological change and substitution that reduce resource inputs and emissions per unit of output. Significant mitigation can be achieved through improved cropland and grazing land management, restoration of degraded lands, and land use change like agroforestry to increase soil carbon storage. Emissions from livestock production can be reduced through improved nutrition and better management of manure. Improved land management through mulching, minimum/zero tillage, intensive cropping, growing legumes, green manuring, crop residues management, improved nutrient management, improved composting including vermicomposting can be undertaken to mitigate the climate change impact on agriculture. A large proportion of the mitigation potential of agriculture arises from soil carbon sequestration, which has strong synergies with sustainable agriculture and generally reduces vulnerability to climate change. Improved rice cultivation techniques can reduce methane gas emissions. Improved nitrogen fertiliser application techniques can also reduce N_2O emissions. Suitable energy crops should be encouraged to replace fossil fuel use.

The key mitigation technologies in agriculture, as per National Mission for Sustainable Agriculture (NMSA, 2010) are

1. Improved crop seeds, livestock and fish cultures
   - Promoting use of biotechnology
   - Research and promotion of C4 pathways in C3 plants
   - Conserving indigenous genetic resources
   - PPP (Public-Private Partnership) in R&D, management and dissemination of improved varieties
   - Conserving agricultural heritage

2. Water use efficiency
   - Promoting water use efficiency in agriculture
   - R&D in the areas of energy efficient water system
   - Developing mechanism for integrated management of rainwater, surface and groundwater
   - Policy instruments for PPP
   - Strengthening local institutions in managing water allocation and utilization

3. Pest management
   - Efficient, safe and environmentally sound methods of pest management
   - Research, commercial production and marketing of biopesticides
   - Developing insect forecasting models
   - Decision and information support system for pest and disease surveillance
   - Institutional mechanism for quick response in case of disaster

4. Improved agronomic practices
   - Improved agronomic practices to reduce farm losses
   - Conservation and precision farming

- Knowledge management
- Soil conservation
- Policy instruments for optimum land use

5. Nutrient management
   - Production and use of biofertilizers
   - Developing nutritional strategies for managing heat stress in dairy animals
   - Strengthening capacity of existing soil testing labs
   - Quality standards and quality control systems for raising confidence among users

6. Agricultural insurance
   - Developing various models for risk management
   - Designing users' friendly DSS (Decision Support System) to assess risks and develop contingent plans
   - Weather based crop insurance schemes
   - Region-specific contingency plans based on vulnerability and risk scenerio

7. Credit support
   - Developing new forms of credit assessment and risk management system
   - Promoting micro-finance
   - Developing mechanisms to enhance the flow of credit to critical infrastructure
   - Up-scaling kisan credit card scheme
   - Designing customised credit policies and programmes to mitigate risks

8. Markets
   - To formulate market-aligned R&D programmes
   - Improving supply chain efficiency
   - Creation of new market infrastructure
   - Supporting community partnerships in developing food and forage banks
   - Strengthening access to quality and timely inputs by farmers

9. Access to information
   - Minimising information asymmetry through information and communication technology-based system

- PPP to develop technology based solution to market related problems
- Building an ICT (Information and Communication Technology) enabled knowledge management network
- To create, manage and develop national resource portal

10. Livelihood diversification
    - Mitigating risks by supplementing income from off-farm activities
    - Crop diversification
    - Crop-livestock-fisheries farming system

REDUCING IMPACTS OF CLIMATE ON LIVESTOCK PRODUCTION

For an animal to maintain homeothermy (no change in body temperature, other than normal circadian rhythms), the ambient environment and the animal must exchange heat at a rate that permits balancing the metabolic heat production and the energy gains/losses through the four transfer modes, *i.e.*, radiation, convection, conduction and evaporation. Ruminant animals primarily adjust evaporative heat loss to maintain homeothermy during brief exposures to adverse weather, but will reduce feed intake to lower heat production during prolonged hot weather. Swine and poultry primarily adjust heat production to maintain homeothermy. To alter the microclimate of an animal effectively through housing or environmental modification the following factors should be considered (WMO, 2010).

Site selection

The selection of a site for housing of animals is the most important factor to minimise the effects of local weather. Climatic factors vary with height above the ground at a specific location. Observations of the microclimates in a general location will reveal much variation in thermal conditions resulting from terrain features, differential exposure, wetlands, rivers, type and height of vegetation, human activities and other factors. Proper selection of a site to emphasise factors for enhanced heat dissipation will have long-term protection benefits.

Windbreaks

Windbreaks have an important role to reduce the climatic impacts on animals in tropical and subtropical regions. Grazing animals are protected from strong winds, especially during cold weather. Structures or trees can markedly reduce wind speed and can be beneficial to the survival of exposed animals, especially the newborns. A windbreak acts as a barrier that lowers the wind speed near the ground surface and diverts and splits the air stream. The protection achieved is determined by the configuration, height, density and thickness of the trees in a belt. The use of leguminous trees or shrubs can be a practical means to counteract the effects of wind and heat stress, as well as to improve animal diet.

Shades

Shades may be provided for protecting farm animals in hot climates. In a tropical region, the solar irradiance is high even during the winter. It is reported that when dairy cows are given access to adequate shades, milk production is increased. Enclosed shelters are not recommended for tropical climates because of the decreased natural air velocity and sanitation. In temperate regions, partially enclosed shelters can reduce the thermal radiation received by animals during hot weather.

Genetic improvement for adaptation

Many local breeds are adapted to harsh living conditions and they are able to survive, grow and reproduce in conditions of poor nutrition, and even under infections of parasites and disease causing microorganisms. Any population must be genetically heterogeneous in order to be able to survive under the challenges of the changing environment. Thus, adaptation strategy should be identifying and strengthening local breeds that have adapted to local climatic stress and feed sources and improving local genetics through cross-breeding with heat and disease tolerant breeds. This is the basis for genetic improvement in livestock. If climate change is faster than natural selection, the risk of survival and adaptation of the new breeds is greater (Hoffmann, 2008).

Chapter 15

Agrometeorological Observatory

The science of agrometeorology has great practical utility in protection against, or avoidance of adverse climate risks of crop production. Agrometeorological information can be used in efficient land use planning, determining suitable crops for a region, risk analysis of climate hazards and profit calculations in farming, production and harvest forecasts, and in adoption of farming methods and choice of farm machinery. Meteorological instruments are the equipments used to sample the state of the atmosphere at a given time. Meteorology does not use much laboratory equipments but relies more on field-mode *in situ* observation and remote sensing equipments. Rain was one of the first weather parameter to be measured historically. Two other accurately measured weather-related variables are wind and humidity.

An agrometeorological observatory or station is a place where all the necessary instruments and structures are installed and maintained to observe and record the weather parameters at stipulated time interval.

Table 15.1 Instruments used for measuring weather parameters in observatories and laboratories

Instrument	Weather parameter measured
Maximum thermometers	Highest temperature attained during a day
Minimum thermometers	Lowest temperature attained during a day
Grass minimum thermometer	Minimum temperature near ground surface
Thermograph	Air temperature continuously
Thermocouples	Soil temperature
Soil thermometer	Soil temperature
Aneroid barometer	Atmospheric pressure
Barograph	Atmospheric pressure continuously
Ordinary rain gauge	Amount of rainfall
Self-recording rain gauge	Amount and intensity of rainfall
Psychrometer/ hygrometer	Relative humidity

(*Contd.*)

Instrument	Weather parameter measured
Assmann psychroz	Relative humidity in crop canopy
Hair hygrometer	Relative humidity inside the room
Hygrograph	Relative humidity continuously
Wind vane	Wind direction
Anemometer	Wind velocity
USWB open pan evaporimeter	Evaporation
Heliograph	Duration of bright sunshine hours
Bellani pyranometer	Total incoming radiation
Eppley spectral pyranometer	Photosynthetically active radiation
Spectroradiometer	Solar radiation in narrow wave bands
Kipp radiometer	Diffused radiation
Pyrgeometer	Long wave radiation
Pyradiometer	Both long and shortwave radiation
Net radiometer	Net radiation
Quantum sensor	Quantum content in radiation
Spectrophotometer	Wavelength of light

CLASSIFICATION OF OBSERVATORIES

Agrometeorological observatories in India are classified into three major (A, B and C) and two minor (D and E) classes or types. Class A, B and C observatories are equipped with both eye reading and self recording instruments. A, B and C class observatories are also known as principal, ordinary and auxiliary agrometeorological observatories, respectively. Observations in A class observatories are recorded at an interval of three hours in a day whereas in B and C class observatories observations are recorded thrice and once, respectively in a day.

Principal agrometeorological observatory

A principal agrometeorological or A-type observatory provides detailed simultaneous meteorological and biological information and it is where research in agricultural meteorology is carried out. The instrumental facilities, range and frequency of observations, in both meteorological and biological fields, and the professional personnel are such that fundamental investigations into agricultural meteorological questions of interest to the countries or regions concerned can be carried out (WMO, 2003). In India, the central agromet observatory situated at Pune is one of such observatories. The main activities of this observatory include recording of large number of meteorological parameters of interest to agriculture, coordinating All India

Coordinating Crop Weather Projects on different crops and issue of monthly crop outlooks, preparation and publication of agroclimatic atlases and crop-weather diagrams, and collaborating with World Meteorological Organisation (WMO) in the field of agrometeorology and allied disciplines. Self-recording instruments record most of the data on continuous basis. These are supported and supplemented by spot observations at 0700 and 1400 hour Local Mean Time (LMT).

Ordinary agrometeorological observatory

An ordinary agrometeorological observatory provides, on a routine basis, simultaneous meteorological and biological information and may be equipped to assist in research into specific problems. In general, the programme of biological or phenological observations for research will be related to the local climatic regime of the station as well as to local agriculture (WMO, 2003). In India, there are two types of observatories in this category, crop-weather and non crop-weather observatories. A crop-weather observatory records data on growth and yield of one or more of the major crops according to standard plan along with the meteorological data. The data are sent to Agromet Division of India Meteorological Department (IMD), Pune where these are compiled and crop-weather relationships are drawn. Each NARP agroclimatic zone has one ordinary agrometeorological observatory at its zonal research station.

Auxiliary agrometeorological observatory

An auxiliary agrometeorological station provides meteorological and biological information. The meteorological information may include such items as soil temperature, soil moisture, potential evapotranspiration, duration of vegetative wetting, and detailed measurements in the very lowest layer of the atmosphere. The biological information may cover phenology, onset and spread of plant diseases, and so forth (WMO, 2003). These observatories are usually established in agricultural farms and are equipped with few instruments to collect qualitative data.

Minor observatories

An agrometeorological station for specific purposes or a D-type observatory is a station set up temporarily or permanently for the observation of one or several variables and/or specified phenomena. The 'pilot balloon cum micrometre stations' are used under desert locust meteorology scheme to assist in the issue of anticipated wind for locust observations. The IMD uses E-type or 'evapotranspiration stations' to measure evapotranspiration by using lysimeter.

INSTRUMENTS FOR AGROMETEOROLOGICAL OBSERVATORIES

Different instruments are used for measuring weather parameters in observatories and laboratories.

Table 15.2 Essential and optional instruments meant for different type of agrometeorological observatory

Principal observatory	Ordinary observatory	Auxiliary observatory
Essential instruments		
Wet and dry bulb thermometers	Wet and dry bulb thermometers	Wet and dry bulb thermometers
Maximum and minimum thermometers	Maximum and minimum thermometers	Maximum and minimum thermometers
Grass minimum thermometer	Grass minimum thermometer	
Soil thermometers	Soil thermometers	
Ordinary rain gauge	Ordinary rain gauge	Ordinary rain gauge
Self-recording rain gauge		
Wind vane	Wind vane	
Anemometer	Anemometer	
USDA open pan evaporimeter	USDA open pan evaporimeter	
Sunshine recorder		
Assmann psychrometer	Assmann psychrometer	
Dew gauge		
Thermohygrograph		
Soil moisture equipments		
Soil radiation instruments		
Optional instruments		
Lysimeter	Sunshine recorder	Wind vane
Potentiometer	Self-recording raingauge	Anemometer
Microvoltmeter	Dew gauge	Dew gauge
Thermopile	Thermohygrograph	

LAYOUT PLAN OF AN AGROMETEOROLOGICAL OBSERVATORY

Ideal site for an observatory is a well exposed and levelled plot of size 55 m (N-S) × 36 m (E-W) located more or less at the centre of the farm. To ensure that the observations are representative of the region and comparable with those made at other observatories, the following criteria should be adhered to while selecting a site for an agrometeorological observatory.

1. The observatory site should be truly representative of the agricultural region concerned.
2. The observatory, as far as possible should be located at the centre of the farm.
3. The site of the agrometeorological observatory should be fairly levelled without any slope and free from obstructions.
4. Obstructions such as trees, buildings and nearby shrubs should not be closer to the instruments than a distance measuring ten times their height.
5. No obstructions should cast shadows during the greater part of the day.
6. It should have a grass cover and weeds should be removed and grass in and around the enclosure should be frequently mown to keep it uniformly short.
7. Accessibility to the observatory should be another criterion for selection of site from the point of view of both taking observations and maintenance.
8. The site should be away from water bodies or irrigation channels.

The observatory site should be enclosed with barbed wire fence up to 1.2 m height to avoid trespasses by animals and people. There should be a gate at appropriate location. Usually the agrometeorological observatories are set up in agricultural universities and research farms under the technical guidance of the IMD, Pune. Technical personnel from the IMD visit the observatory from time to time for inspection.

All tall instruments like sunshine recorder, wind vane, anemometer, *etc.* should be installed on the northern side of the observatory to avoid shade effect. Some instruments are to be installed at prescribed height in the observatory. The IMD has also prescribed specific size and diameter, *etc.* for the instruments.

Table 15.3 Height of instruments from ground level installed in the observatory

Instrument	Height of instruments from ground level (m)
Microclimatic post	3.68
Sunshine recorder	3.05
Anemometer	3.05
Wind vane	3.05
Stevenson screen	1.22
Ordinary rain gauge	0.30
Self-recording rain gauge	0.75
Evaporation pan	0.30

Table 15.4 Specifications of diameter of the instruments for the observatory

Instrument	Diameter (cm)
Glass globe of sunshine recorder	10.0
Funnel rim of ordinary rain gauge	12.7
Funnel rim of self-recording rain gauge	20.3
Class A open pan evaporation pan	120.7

Time of observation

The surface of earth is divided into 24 'time zones', the way in which there are 24 hours in a day. The time established in each zone is 'standard time'. The local mean time (LMT) corresponding to Indian Standard Time (IST) varies from place to place depending upon longitude of a place. The IST is the LMT for longitude of 82° 30' E (or 82.5° E), which passes through Allahabad. The LMT is based on the transit of the mean sun. To calculate LMT from IST, it is essential to know the longitude of the place. As the earth takes approximately 4 minutes to cross one degree longitude, the time difference between the LMT of a place in India and IST is estimated by multiplying 4 with the difference of longitude of that place and that of Allahabad (82.5° E). If the longitude of the place is more than 82.5° E, the LMT is ahead of the IST and if less than 82.5° E it is behind of the IST.

The observations are recorded at 0700 and 1400 hours LMT all over India. However, rainfall and evaporation observations are recorded at 0830 hours IST and 1400 hours LMT. The settings of automatic instruments like thermograph, hydrograph, evaporigraph and barograph, *etc.* are done at 0830 hours IST.

The observing programme at agrometeorological observatories should include observations of some or all of the following variables characterising the physical environment; solar radiation, sunshine and cloudiness, air and soil temperature, air pressure, wind speed and direction, air humidity and soil moisture, evaporation and precipitation. The WMO has recommended a minimum accuracy for the different variables as given in table 15.5.

Table 15.5 Minimum accuracy recommended by WMO for some variables

Variable	Accuracy required in daily values
Temperature, including max/min, wet and dry bulb, soil	≤ ±0.5° C
Rainfall	±1 mm
Solar radiation including sunshine	10% (±1h)
Evaporation	±1 mm
Relative humidity	±5%
Photoperiod	10% (±1h)

Variable	Accuracy required in daily values
Wind speed	±0.5 m/s
Air pressure	±0.1 hPa

MEASUREMENT OF TEMPERATURE

Air temperature is measured by thermometers housed in a special wooden box called Stevenson screen fixed at 1.22 m above ground level in an observatory. A Stevenson screen is an enclosure to shield meteorological instruments against precipitation and direct heat radiation from outside sources, while still allowing air to circulate freely around them. It forms part of a standard weather observatory. It was designed by Thomas Stevenson, a British civil engineer in 1864. The Stevenson screen is a wooden rectangular box of length 56 cm, width 30 cm and height 40 cm with a double roof having lowered sides. The unit is either supported by four metal or wooden legs or a wooden post. The screen is set up with its door facing north, opening downward to minimise the sunlight entry at the time of recording observations. The whole screen is painted with several coats of white to reflect sunlight radiation and usually requires repainting every two years.

Maximum and minimum thermometers are placed in horizontal position in the wooden box on the upper and lower sides, respectively and bulb-end rest at an angle of 2 degree to horizontal plane. Dry bulb and wet bulb thermometers are kept vertical in the wooden box on the left and right sides, respectively.

Fig 15.1 Stevenson screen

Maximum and minimum thermometers

These are the instrument which show the maximum and minimum temperature of air of the day. Maximum thermometer is a mercury-in-glass thermometer having a constriction near the bulb end. When the temperature of air rises, the mercury in the thermometer expands and forces its way into the stem past this constriction. But when the bulb cools the mercury above the constriction will not get back into the bulb and the length of the mercury thread remains the same. The end of the mercury thread farthest from the bulb registers the highest temperature reached in a day. The temperature range of maximum thermometer is from –35 to +55° C.

The liquid used in minimum thermometer is alcohol in which a dumb-bell-shaped index is set. When the temperature falls, the alcohol column drags the index towards the bulb end but when the temperature rises the alcohol column expands and runs past the index without disturbing it. Thus the end of the index farthest from the bulb gives the lowest temperature attained in a day. The temperature range of minimum thermometer is from –40 to +50° C.

Thermograph

This is an automatic self recording thermometer placed in the Stevenson screen which marks the prevailing temperature continuously on graph paper wounded round a drum operated by clockwork mechanism. The drum makes one revolution in a day and the automatically marked chart gives the temperature variations taking place during the day. The graph paper is changed every day at 0830 hours IST.

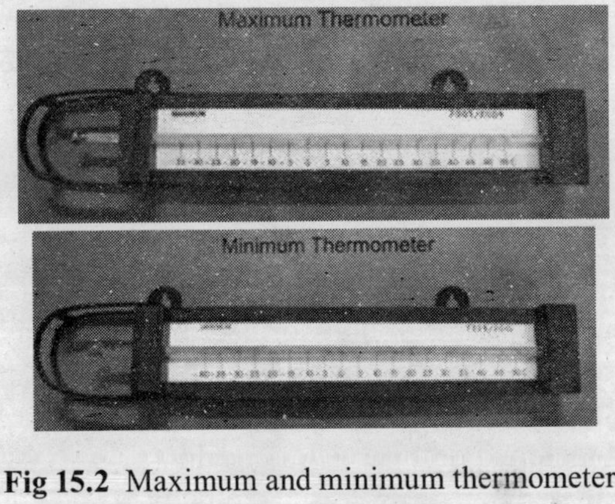

Fig 15.2 Maximum and minimum thermometers

Grass minimum thermometer

Grass minimum thermometer or terrestrial radiation thermometer is used to measure the actual minimum temperature experienced by the plants near the ground surface. It is a glass thermometer filled with alcohol with a small index just like in case of minimum thermometer. Its construction and working is similar to that of minimum thermometer except that its bulb is spherical and the stem is encased in a glasss jacket in order to protect seal marks on the stem from dew, rain and prevent the stem from cooling rapidly. The thermometer is exposed on short grass about 2.5 cm above ground surface on its support over the grass plot on a Y-shaped stand, so that the bulb just touches the grasses. Its reading is oberved before sunrise in the morning. After reading, the thermometer is kept indoors to avoid direct exposure to solar radiation. The grass minimum temperature remains lower than the minimum air temperature at screen level.

Fig 15.3 Grass minimum thermometers

Soil thermometer

Soil temperature directly influences crop growth because the sown seeds, plant roots and microorganisms live in the soil. Thus, agrometeorological observatories should also include soil temperature measurements. Usually soil temperature is measured at 5, 10, 20 and 50 cm depths by mercury-in glass thermometers. These thermometers have a bend of about 120 degree just above the bulb and the rest of the stem is straight. The range of the temperature is –20° to +60° C. Soil thermometers are installed at a small plot of 180 cm × 120 cm in the observatory. These are held by iron stands on inclined position making 60 degree angles with ground surface. Soil thermometers for 5, 10 and 20 cm depths are placed at 45 cm apart at an inclined depth of 5.8, 11.6 and 23.2 cm to ensure a vertical depth of 5, 10 and 20 cm, respectively. The soil temperature is recorded daily at 0700 and 1400 hour LMT. Care should be taken to ensure that good contact is maintained between the thermometer and the soil.

Dry bulb and wet bulb thermometers

Dry bulb thermometer is essentially a mercury-in-glass thermometer having a round bulb at one end and the stem is graduated in degree Celsius. The reading of this thermometer gives the free air ambient temperature at a place when exposed at a height of 1 m to 1.5 m inside a louvered thermometer screen. Temperature of cool air is measured by wet bulb thermometer. A wet bulb thermometer is similar to dry bulb thermometer except that the bulb of the wet bulb thermometer acts as evaporating surface. The bulb is covered by a muslin cloth and is always kept wet by providing water by means of four stands of cotton thread dipped into a small bottle with distilled water. The temperature readings of both dry and wet bulb thermometers will be same under saturated conditions. But when the air becomes dry, the difference would increase. This difference between the dry bulb and wet bulb thermometer readings is called as wet bulb depression, which is used to measure dew point, vapour pressure and humidity.

MEASUREMENT OF RELATIVE HUMIDITY

Relative humidity can be measured by an instrument called a hygrometer. Humidity measurement instruments usually rely on measurements of some other quantity such as temperature, pressure, mass or a mechanical or electrical change in a substance as moisture is absorbed. By calibration and calculation, these measured quantities can lead to a measurement of humidity. Modern electronic devices use temperature of condensation (the dew point), or changes in electrical capacitance or resistance to measure humidity differences. Hygrometers are often called humidity sensors, or relative humidity indicators, which makes their purpose easier to understand. The first practical hygrometer was invented by Johann Heinrich Lambert in 1755.

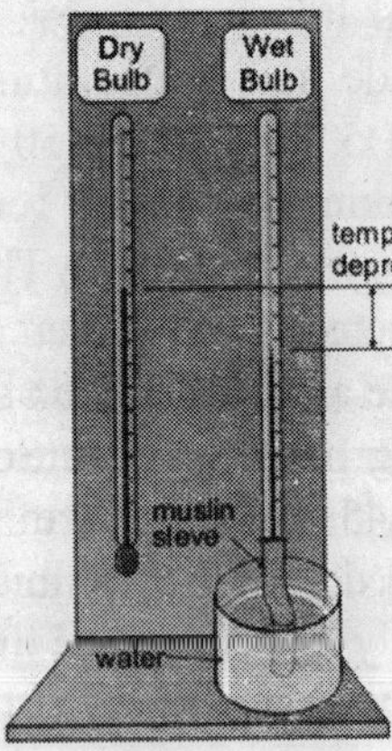

Fig 15.4 Dry bulb and wet bulb thermometers

Psychrometer

The simplest hygrometer is a stationary psychrometer which consists of two thermometers mounted together with a handle attached on a chain and housed vertically in the Stevenson screen. One thermometer is ordinary, called dry bulb thermometer and the other one is a wet bulb thermometer, in which the bulb is always kept moist with distilled water on a sock or wick. Thus a set of dry and wet bulb thermometers is known as a psychrometer. At temperatures above the freezing point of water, evaporation of water from the wick lowers the temperature, so that the wet bulb thermometer usually shows a lower temperature than that of the dry bulb thermometer. When the air temperature is below freezing point, however, the wet bulb is covered with a thin coating of ice and may be warmer than the dry bulb. Relative humidity is computed from the ambient temperature as shown by the dry bulb thermometer and the difference in temperatures as shown by the wet bulb and dry bulb thermometers. Hygrometric table may be used to find out dew point temperature and relative humidity corresponding to dry and wet bulb temperatures.

Whirling psychrometer

The whirling psychrometer is an instrument for the measurement of the temperature and humidity of the air in open and in the crop canopy at different heights. It consists of one dry bulb and one wet bulb thermometer exposed side by side horizontally to a rectangular wooden frame. The thermometer mounted frame can be rotated with a handle. As the instrument is whirled, water evaporates from the container attached to the wet bulb thermometer and cools the thermometer. Four rotations per second should be given to obtain the desired wind speed of about 5 m per second. From the readings of dry bulb and wet bulb thermometers, the dew point temperature, relative humidity and vapour pressure can be calculated using appropriate formulae.

Assman psychrometer

This is an aspirated type portable hygrometer for the accurate measurement of atmospheric humidity. This was devised by Adolph Richard Assmann in 1887. It consists of two mercury thermometers exposed side by side in a steady airflow produced by an attached aspirator. One of the thermometers serves as a dry bulb while the other, with its bulb kept moist with wet sleeve acts as a wet bulb. As water evaporates, the temperature of wet bulb is lowered to an equilibrium point. This depression of wet bulb temperature along with the ambient dry bulb temperature is a measure of moistness or humidity of the ambient air which can be readily obtained from a computed table or slide rule specially made to express in units of relative humidity, dew point temperature or vapour pressure.

Hairline hygrometer

In this type of hygrometer one or more human hairs are held under slight tension by a spring which expands or contracts with changes in the moisture of the surrounding air. It is circular in shape and indicates humidity directly on a dial or scale. It is ideal for laboratory for measuring humidity at room temperature. This is also known as hair hygrometer or hair tension hygrometer.

Hair hygrograph

The hair hygrograph is an instrument used for recording continuously the relative humidity of the air. The sensor of this instrument is a bunch of human hair tied at both ends and kept tight in the middle by means of a hook attached to one arm of the lever and second arm is associated with pen arrangement which can make marking on graph paper attached on the clock driven revolving drum. The length of the hair increases as the humidity increases and *vice versa.* These small changes in length are magnified by means of the lever system which is communicated by second arm of the lever to record the changes on graph paper. The movement of the pen on the graph paper is proportional to the relative humidity. The circular drum makes one rotation in 24 hours or once in a week. The hair hygrograph is kept in a Stevenson screen in an observatory.

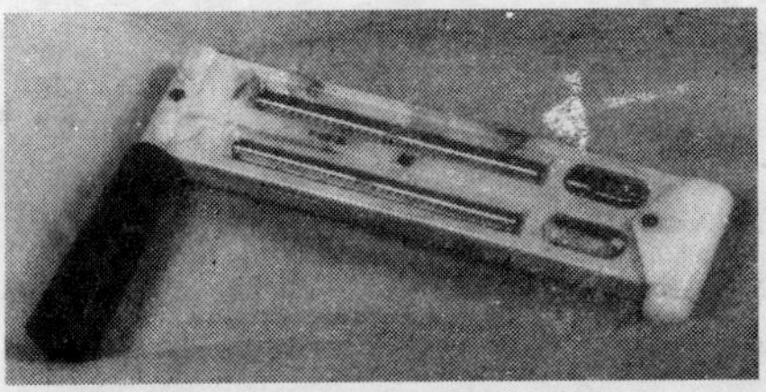

(a) Whirling psychrometer

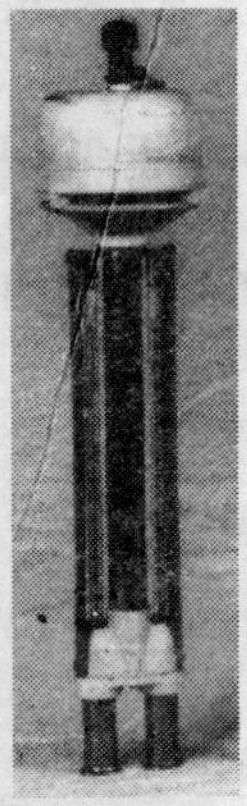

(b) Assman psychrometer

(c) Hair hygrometer

(d) Hair hygrograph

(e) Gravimetric hygrometer

Fig 15.5

Table 15.6 Dry bulb temperature and wet bulb depression with corresponding relative humidity (%)

Relative humidity (%)																
Dry bulb temp (°C)	Difference between wet bulb and dry bulb temperatures (°C)															
	0	**1**	**2**	**3**	**4**	**5**	**6**	**7**	**8**	**9**	**10**	**11**	**12**	**13**	**14**	**15**
-2	100	79	58	37	20	1										
0	100	81	63	45	28	11										
2	100	83	67	51	36	20	6									
4	100	85	70	56	42	27	14									
6	100	86	72	59	46	35	22	10								
8	100	87	74	62	51	39	28	17	6							
10	100	88	76	65	54	43	33	24	13	4						
12	100	88	78	67	57	48	38	28	19	10	2					
14	100	89	79	69	60	50	41	33	25	16	8	1				
16	100	90	80	71	62	54	45	37	29	21	14	7	1			
18	100	91	81	72	64	56	48	40	33	26	19	12	6			
20	100	92	82	74	66	58	51	44	36	30	23	17	11	5		
22	100	92	83	75	68	60	53	46	40	33	27	21	15	10	4	
24	100	92	84	76	69	62	55	49	42	36	30	25	20	14	9	4
26	100	92	85	77	70	64	57	51	45	39	34	28	23	18	13	9
28	100	93	86	78	71	65	59	53	47	42	36	31	26	21	17	12
30	100	93	86	79	72	66	61	55	49	44	39	34	29	25	20	16

Modern hygrometers

Now electronic hygrometers are available which can give humidity reading instantly and electronically. They measure the capacitance or resistance of a sample of air and calculate the humidity from that.

In a capacitive hygrometer, there are two metal plates with air in between them. The more water there is in the air, the more it affects the plates' capacitance (ability to store a static electric charge). By measuring how much charge can be stored, it is possible to measure the humidity quickly and accurately.

In a resistive sensor, electricity flows through a piece of ceramic material exposed to the air. The higher the humidity, the more water vapour condenses inside the ceramic material, changing its resistance. Measurement of how much current flow through the ceramic material gives an accurate measurement of the humidity. In thermal conductivity humidity sensors, the change in thermal conductivity of air due to humidity is measured.

A gravimetric hygrometer measures the mass of an air sample compared to an equal volume of dry air. This is considered the most accurate primary method to determine the moisture content of the air.

MEASUREMENT OF DIRECTION AND SPEED OF WIND

The measurement of wind direction and speed is one of the most important factors in weather prediction. It plays an important role in determining the crop water use.

Wind vane

Wind vane is an instrument that indicates the direction of wind. It consists essentially of a brass-arm, placed in ball bearings to a vertical axis to enable the arm to move freely in the horizontal plane. To one side of the brass arm there is a broad arrow head and on the other side there is a flat vane. Below the vane there are four direction arms fixed to the vertical axis. The direction arms are labelled with N, S, E and W. The sensitive part of the instrument is the flat vane, which offers the greatest resistance to the motion of the air going to the leeward side and thus, the arrow indicates the direction of wind. Wind vane is installed over a wooden plank fixed on a wooden post. The height between the pointer and ground level is 3.05 m. The north indicator should be set to geographical north and not to the magnetic north.

Anemometer

An anemometer is a device used for measuring wind speed and it is a common weather station instrument. The most common type anemometer is the Robinson cup anemometer. In 1846, John Thomas Romney Robinson designed this by using four hemispherical cups and mechanical wheels. Later, in 1926, John Patterson developed a three cup anemometer.

In an anemometer either three or four metal hemispheres are mounted on arms which rotate freely about a vertical axis. A system of gears similar to an automobile speedometer translates the rotation generated by the wind to an indicator or recorder for the speed of wind that has passed the instrument. The anemometer is installed on a metal pipe, which is fixed on a wooden post. The height from centre of the anemometer cups is 3.05 m above ground level.

One of the other forms of mechanical velocity anemometer is the vane anemometer. It may be described as a windmill or a propeller anemometer. Contrary to the Robinson anemometer, where the axis of rotation is vertical, the axis on the vane anemometer is parallel to the direction of the wind and therefore horizontal. Since the wind varies in direction and the axis has to follow its changes, a wind vane is also employed to the instrument. A vane anemometer thus combines a propeller and a tail on the same axis to obtain accurate and precise wind speed and wind direction measurements from the same instrument.

Now-a-days digital anemometers are also used for the measurement of wind speed. Digital anemometer gives more precise and accurate reading.

(a) Wind vane

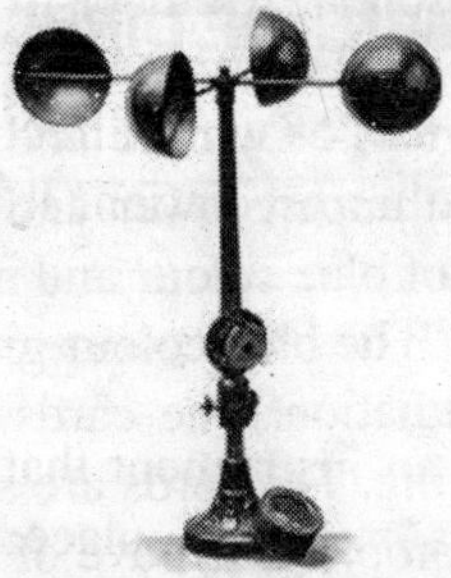

(b) Robinson anemometer

(c) Vane anemometer

Fig 15.6

MEASUREMENT OF BRIGHT SUNSHINE HOUR AND SOLAR RADIATION

Plants get their required energy for life from sunlight. They convert solar energy into chemical energy during the process of photosynthesis. Time of flowering of many plants is influenced by duration of bright sunshine or photoperiod. Thus crop production is the exploitation of solar radiation.

Sunshine recorder

Continuous record of bright sunshine hour is obtained by a sunshine recorder. The commonly used instrument is Campbell-Stroke sunshine recorder. It is installed on a masonry pillar of 3.05 m above the ground. The instrument consists of a spherical glass of 10 cm diameter. This glass sphere is mounted concentrically in a section of a spherical bowl in such a way that the sun's rays are focussed sharply on a card held in the groove cut into the bowl. Three overlapping pairs of grooves are provided in the bowl to take the cards. There are three different types of sunshine cards namely the short curved card for winter season (15 October to end of February), the long curved card for summer season (12 April to 2 September) and the straight card for equinoxes (3 September to 14 October and 1 March to 11 April) and these cards are inserted in the upper, lower and middle grooves, respectively in the bowl. The cards are of blue colour and made up of good quality pasteboards of 0.04 mm thickness. The blue colour gives good contrast with the burns and freely absorbs the radiation. The cards are so cut that no expansion occurs in its length on wetting. The cards are subdivided into hourly intervals. The card is inserted in appropriate groove of the recorder in such a way that the 12 hour line coincides with the noon mark engraved on the bowl and the clamping screws are tightened accordingly. As the sun moves across the sky, its focussed image burns a trace of the card. The burns are then measured with a sunshine scale, specially made for the purpose. The scale is made up of chemically sensitised cellulose paper and each hour is divided into 10 parts consisting 0.1 hour or 6 minutes. The parallel sunshine scale is used for straight cards and trapezoidal scale is used for long and short curved cards. The card is changed every day and is inserted before sunrise and the burnt card is removed after sunset. The card burns linearly due to chemical treatment given to it. Total length of the burn is measured by adding up the values of all the hours and total duration of sunshine hours for the day is determined. The hours marked on the card indicate local mean time of the station and sunshine is measured in number of hours per day.

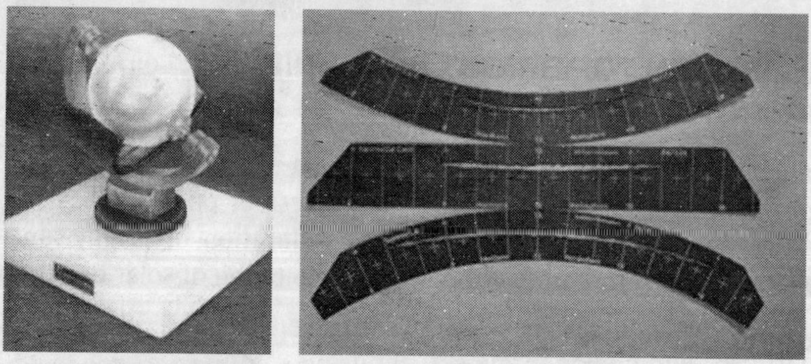

(a) Sunshine recorder (b) Sunshine cards

Fig 15.7

Actinometers

Actinometers or radiometers are instruments used to measure the heating power of radiation. In meteorology, they are usually oriented to measure downwelling or upwelling radiation. An actinometer is a chemical system or physical device in which sensors determine the number of photons in a beam integrally or per unit time.

Insolation is the total amount of solar radiation received on a given surface area during a given time. It is also called solar irradiation and expressed as 'hourly irradiation' if recorded during an hour or 'daily irradiation' if recorded during a day. The unit recommended by the World Meteorological Organisation is the megajoule per square metre (MJ/m^2) or joule per square millimetre (J/mm^2). An alternate unit of measure is the Langley (1 thermochemical calorie per square centimetre or 41,840 J/m^2). If this energy is divided by the recording time in hours, it is then a density of power called irradiance, expressed in watts per square metre (W/m^2).

Special categories of actinometers are designed to measure different forms of solar radiation like direct solar radiation, diffused radiation, reflected radiation, thermal radiation and net radiation. Depending upon the requirement, different instruments are used for recording the data.

A pyranometer is a type of actinometer used to measure broadband solar irradiance and is a sensor that is designed to measure the solar radiation flux density from a field of view of 180 degrees. A pyranometer's main components include a thermopile sensor with a black coating and a glass dome. The thermopile sensor absorbs all solar radiation and it has a flat spectrum covering the 300 to 50,000 nanometre range. The glass dome limits the spectral response from 300 to 2,800 nanometers (cutting off the part above 2,800 nm), while preserving the 180 degrees field of view. Another function of the dome is that it shields the thermopile sensor from convection. The black coating on the thermopile sensor absorbs the solar radiation. This radiation is converted to heat. The heat flows through the sensor to the pyranometer housing. The thermopile sensor generates a voltage output signal that is proportional to the solar radiation. There are several types of pyranometers like eppley pyranometer, bellani pyranometer, precision pyranometer and albedometer.

A pyranometer that measures only diffuse solar radiation scattered from air, particles, and clouds in the sky, by using a device that shades the sensor from direct sunlight is called diffusometer. Pyrgeometer is an instrument that measures infrared (longwave) radiation from a hemisphere that strikes a flat, horizontal surface.

Quantum sensor is used for measuring photosynthetically active radiation (PAR). Tube solarimeters containing several quantum sensors are used for measuring the intercepted PAR in the crop canopy and the quantum sensors for measuring PAR near the ground. Difference between the two gives the intercepted or absorbed PAR.

Sensors that measure the difference between down- and up-welling radiations at all wavelengths are called net radiometers. In a net radiometer there are two sensors, which are exposed to the sky and earth. Sensor exposed to sky measures incoming radiation and the other facing the earth measures outgoing radiation.

Solar radiation in narrow wave bands can be measured by spectroradiometer. It measures radiation as a function of wavelength, to determine the spectrum of radiation. Relationship between plant morphology and reflectance can be measured with this instrument.

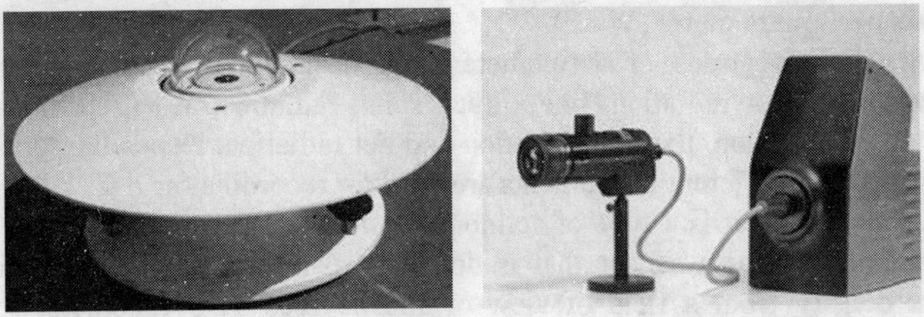

(a) Pyranometer (b) Spectroradiometer

Fig 15.8

MEASUREMENT OF RAINFALL

Water in any form falling on the earth is called precipitation. Rain, the liquid form of precipitation is the main source of soil moisture for crop production. Rainfall is measured as the amount of water that reaches horizontal ground or the horizontal ground projection plane of the earth's surface, and is expressed as a vertical depth of water. Rain gauge is a device used for measuring the amount of rainfall Rain gauge may be a non recording or self recording type.

Non recording rain gauge

The essential parts of an ordinary or non recording rain gauge are a funnel through which the rain water is collected in a receiver and a measuring glass with which the rain water collected is measured. There is an additional metallic cylinder that holds both funnel and receiver in order to avoid possible losses due to overflow from the receiver. The correct type of measuring glass suitable to the type of rain gauge in use has to be used for the measurement of rainfall. The rain measure is graduated for every 0.2 mm of rainfall.

The rain gauge is installed at a standard height of 30 cm above the ground level. The 20 cm capacity rain gauge having the receiver with an orifice area of 200 cm^2 and 4 litres bottle is widely used in most of Indian observatories and is sufficient to measure 24 hours rainfall. The 40 and 100 cm capacity rain gauges are used in the observatories located in the high rainfall areas. Rainfall collected in the receiver is measured daily at 0830 IST hour and 1400 LMT hour, if rain occurs. The process of measurement at 0830 IST hour for the past 24 hours is most common throughout the country. However, at times of heavy rainfall, two or three intermediate readings may be taken and summed up to get the rainfall for the past 24 hours. Rainfall exceeding 2.5 cm in a day is called a rainy day.

Self recording rain gauge

Self recording rain gauges give a permanent automatic record of rainfall without any bottle reading. It has a mechanical arrangement by which the total amount of rainfall since the start of record gets automatically recorded on a graph paper. These rain gauges are also called integrating rain gauges since they record cumulative rainfall. In addition to the total amount of rainfall at a station, it gives the times of onset and cessation of rains. They can provide continuous record for a number of days. They are very useful in hilly and far off areas. In other areas, they are installed along with a non recording rain gauge. Various models have been designed with different gauges such as floating type, weighing type and tipping bucket type.

A float type gauge, provided with a self starting siphoning arrangement, is most widely used in India and popularly called as natural siphon recording rain gauge. It consists of a funnel-shaped collector at the top of the gauge and a float siphon chamber and recording mechanism just below it. The rain water collected by the collector is led into the float chamber. The float is having a central stem on which a pen assembly is fixed. The pen is moving over a chart wound over a clockwise rotating drum. The rotating drum completes one revolution in 24 hours or sometimes in 7 days and the chart has to be replaced accordingly. After every one centimetre of rainfall siphoning occurs and recording of rain starts afresh. If there is no rainfall, the pen traces a horizontal line.

The tipping bucket type rain gauge generates an electric signal or pulse for each unit of rainfall collected, and allows automatic or remote observation with a recorder or a counter. It measures the rainfall with a least count of 1 mm and gives out one electrical pulse for every mm of rainfall. Rain gauge consists of twin buckets mounted on a horizontal shaft. It has two stable positions. The collector area of the rain gauge is 324 cm^2. Thus a volume of 32.4 cm^3 corresponds to 1 mm of rainfall. The rainfall collected by the receiver reaches to one of the twin bucket through a funnel. The weight of 32.4 cm^3

of rainfall causes the tilting of the bucket from one stable position to another stable position and brings the other bucket under the funnel for collecting water. Every time the bucket tilts, the small magnet attached to the horizontal shaft of the bucket actuates an electrically driven pen to trace a record on the graph paper mounted on a clockwise rotating drum.

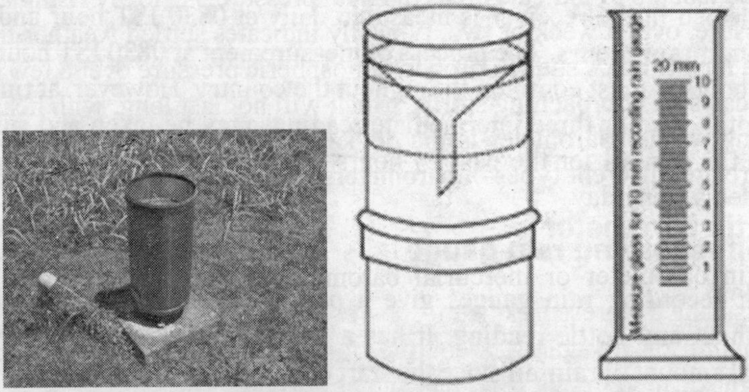

(a) Ordinary rain gauge

(b) Self-recording rain gauge

Fig 15.9

MEASUREMENT OF DEW

Dew is measured by Duvdevani dew gauge and a dew album. This instrument consists of rectangular wooden plates of 32 × 5 × 2.5 cm^3 in dimension, coated with a special red paint. The dew gauge is exposed on a stand in the open around sunset. For measuring the dew formation at different heights, dew gauges are fixed at 5, 25, 50 and 100 cm above ground level. Before sunrise the readings are taken. The size, form and distribution of dew drops on the surface of the plates are compared with a set of photographs in the dew album and a dew scale number is given for the amount of dew present on the surface of the plates.

MEASUREMENT OF ATMOSPHERIC PRESSURE

Changes in atmospheric pressure are one of the most commonly used ways to forecast changes in the weather because weather patterns are carried around in regions of high and low pressure. Weather maps use lines of equal pressure called isobars to indicate areas of equal pressure. A slowly rising atmospheric pressure, over a week or two, typically indicates settled weather that will last for a long time. A sudden drop in atmospheric pressure over a few hours often forecasts an approaching storm, which will not last long, with heavy rain and strong winds. Barometer is the device that measures atmospheric pressure. There are different types of barometers.

Fortin barometer

Fortin barometer or mercurial barometer is the most accurate device for measuring the pressure of atmosphere. It operates on the principle of balancing a column of air against a column of mercury in a sealed glass tube. If the weight of mercury is less than the atmospheric pressure, the mercury level in the glass tube rises. If the weight of mercury is more than the atmospheric pressure, the mercury level falls. Atmospheric pressure is basically the weight of air in the atmosphere above the reservoir, so the level of mercury continues to change until the weight of mercury in the glass tube is exactly equal to the weight of air above the reservoir.

Fortin barometers use a variable displacement mercury cistern, usually constructed with a thumbscrew pressing on a leather diaphragm bottom. The mercury level can be raised or lowered with by the screw. In the cistern vessel, a glass tube filled with mercury, containing an ivory pointer, is kept inverted. The zero of the scale starts from lower tip of the pointer. The level of mercury in the cistern vessel is set to the zero level before the pressure is read on the column. Scales in both centimetres and inches are available on two sides of the instrument along with vernier callipers for accurate reading. To take the reading, height of mercury column is measured on the main scale followed by vernier scale reading. Atmospheric pressure is the sum of main scale reading, and vernier scale reading multiplied by vernier constant. Atmospheric pressure is expressed commonly by the height of the column of mercury in the barometer, either in inches, or millimetres. One atmosphere (1 atm) is equivalent to 760 mm of mercury.

Expansion of metal scale and mercury are different at different temperatures. As such, they are transformed to one common temperature which is zero degrees Celsius. The gravitational pull varies with latitude, requiring corrections for reading. All readings are transformed into common latitude, *i.e.*, 45° N. The readings also need correction to sea level height. Thus, all the three corrections, such as temperature, gravity and latitude are to be applied for reporting the observations.

Aneroid barometer

An aneroid barometer is an instrument for measuring pressure as a method that does not involve liquid. This device was invented in 1844 by the French scientist Lucien Vidi. The aneroid barometer uses a small, flexible metal box called an aneroid cell or capsule, which is made from an alloy of beryllium and copper. The evacuated capsule is prevented from collapsing by a strong spring. Small changes in external air pressure cause the cell to expand or contract. This expansion and contraction drives mechanical levers such that the tiny movements of the cell are amplified and communicated through chain and pulley to the pointer, which moves on graduated scale. The pointer gives direct pressure reading.

Barograph

Barograph is an instrument for the continuous recording of atmospheric pressure exerted at a given place. It is a recording aneroid barometer where the pointer is replaced with a scribe or pen. A barograph uses one or more aneroid cells acting through a gear or lever train to drive a recording arm that has at its extreme end either a scribe or a pen. A scribe records on smoked foil while a pen records on paper using ink. The recording material is mounted on a cylindrical clockwise rotating drum. Commonly, the drum makes one revolution per day, per week, or per month and the rotation rate can often be selected by the user. The paper or foil chart is called a barogram that records the barometric pressure over time.

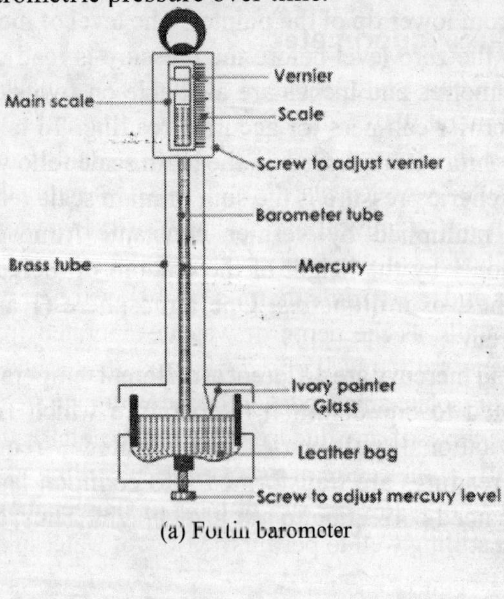

(a) Fortin barometer

(b) Aneroid barometer'

(c) Barograph

Fig 15.10

MEASUREMENT OF EVAPORATION

Evaporation is one of the important aspects of hydrologic cycle by which water is transferred from the earth's surface to atmosphere. The rate of evaporation is expressed in depth of water lost per unit time. Evaporation rate from the crop field determines the irrigation water requirements of the crop. Measurement of evaporation is important in agricultural and hydro-meteorological studies for the irrigation and drainage systems, and for the design and operations of water reservoirs. There are several methods for measuring evaporation such as pan evaporation method, empirical formulae, water balance method and energy budget method. The United States Weather Bureau Class A open pan evaporimeter or USWB class A pan is most commonly used in the observatories of India.

USWB Class A open pan evaporimeter

The Class A evaporation pan consists of a cylindrical reservoir made up of galvanised iron or moulded metal with a diameter of 120.7 cm and height of 25.4 cm. The bottom of the pan is raised 15 cm above the ground surface by resting it on a carefully levelled wooden platform. The pan is painted white and is covered with a wire mesh to prevent animals and birds from drinking water from it. The depth of water is to be kept in a fixed range such that the water surface is at least 5 cm and never more than 7.5 cm below top of the pan.

Evaporation is measured daily as the depth of water evaporated from the pan with a fixed-point gauge. A stilling well of size 10 cm diameter and 30 cm height is placed in the tank to isolate small portion of the water surface in the tank to avoid disturbance by waves if any due to wind. In the point gauge, the pointed brass is fixed vertically at the centre of the cylindrical stilling well. The tip of the rod is located at 6-7 cm below the rim of the pan. Three small holes are there at the bottom of the stilling well to permit the flow of water in and out of the well. At each observation, water level is brought to the same fixed point (reference point) by adding water with a graduated measuring cylinder, devised for the purpose. The cross-sectional area of the measuring cylinder is exactly one hundredth of the area of evaporation pan. The scale from 0 to 20 cm is engraved inside it along the height and the graduation runs from top to bottom in ascending order. One full cylinder of water raises 2 mm height on the pan. Evaporation can be measured accurately up to 0.1 mm least count.

When there is no rainfall, evaporation rate will be equal to water added to the evaporation pan with the measuring cylinder up to the tip of the pointer of the stilling well. When there is rainfall, but the tip of the fixed point gauge remains above the water level, then evaporation rate is equal to the quantity of rainfall plus the quantity of water added to the evaporation pan up to the reference point. When rainfall is more than evaporation, the water level inside the stilling well will be above the reference point. In this case, water is removed from the reservoir with measuring cylinder till the water level comes down to the reference point and difference between the actual rainfall of the previous day and water removed from the reservoir gives the evaporation rate. If there is heavy rainfall with over flow from the pan, evaporation cannot be recorded. In this case, 'overflow' is written in the weather report.

AUTOMATIC WEATHER STATION

An automatic weather station (AWS) is an automated version of the traditional weather station, either to save human labour or to enable measurements at remote areas. The system may report in several different ways. It may be in real time *via* a local link to a computer system or *via* telecommunications or satellite systems or save the data for later recovery. In the past, automatic weather stations were often placed where electricity and communication lines were available. Nowadays, the solar panel, wind turbine and mobile phone technology have made it possible to have wireless stations that are not connected to the electrical grid or telecommunications network.

An AWS will typically consist of a weather-proof enclosure containing a data logger, data logger terminal, data storage system and the meteorological sensors with an attached solar panel or wind turbine mounted upon a mast. The specific configuration may vary with the purpose of the system. The data logger collects the information of each sensor and archives it. The data logger saves all the data either on its own memory or on a memory card. The data logger is the heart of the AWS. It is usually kept in a nearby room attached to different sensors for weather data collection. It can also be left in the field. The unit has 8 to 12 analog channels, four digital channels, two pulse counters and one frequency-sensing channel. All the sensors for recording different weather parameters are connected to appropriate channels of the field unit. The field unit processes the raw meteorological data into practical units of scientific measures with the data logger. The data logger terminal is used to signify the field unit about which sensor to use, what channels these sensors are connected to, when to store data and how top label and organise the data. Thus, all the instructions to field unit are given through the terminal. It has also the facility to display the data when required. The main power source for an automatic weather station is usually one or more solar panels connected in parallel with a regulator and one or more rechargeable batteries. In the northern hemisphere, the solar panel is mounted facing south and in the southern hemisphere it is faced to north. The output from the solar panels may be supplemented by a wind turbine to provide power during periods of poor sunlight, or by direct connection to the local electrical grid.

Most automatic weather stations have thermometer for measuring temperature, anemometer for measuring wind speed, wind vane for measuring wind direction, hygrometer for measuring humidity and barometer for measuring pressure. Some stations can also have ceilometer for measuring cloud height, rain gauge for measuring liquid-equivalent precipitation, ultrasonic snow depth sensor for measuring depth of snow and pyranometer for measuring solar radiation.

Sensors of AWS

Temperature is sensed by thermistors. Thermistors are thermally sensitive resistors whose prime function is to exhibit a large, predictable and precise change in electrical resistance when subjected to a corresponding small change in ambient temperature. These are constructed from a ceramic type semiconductor material using metal oxide technology such as manganese, cobalt and nickel, *etc.* The semiconductor material is generally formed into small pressed discs or balls which are hermetically sealed to give a relatively fast response to any change in temperature. A probe, which is electrically identical to temperature probe, also measures soil temperature. Air temperature sensors are kept in Stevenson screen and soil temperature probes are buried in soil at desired depth. The relative humidity sensor contains both RH and temperature probe and housed in plate gill radiation shield with a 1.5 m lead length. The shield helps to eliminate radiation loading the sensor and also allows ventilation.

Fig. 15.11 USWB Class A open pan evaporimeter

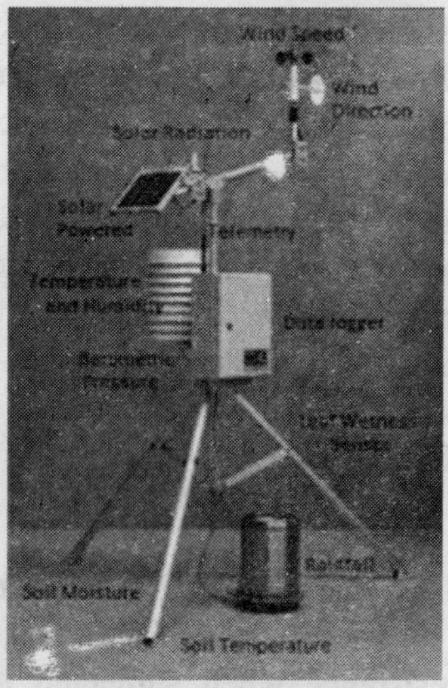

Fig. 15.12 Automatic weather station

Wind vane measures the wind direction from 0 to 360 degree with 5 degree accuracy. The sensor utilises a potentiometer to vary the sensor resistance in relation to wind direction. Anemometer measures wind speed in the range of 0 to 160 km/hour. This sensor is a three-cup wheel assembly utilising a magnet activated reed switch whose frequency is proportional to wind speed. The 014A wind speed sensor and 024A wind direction sensor are suitable for a wide range of wind study applications. They are designed for long term unattended operations in most environments. The sensors use a quick-connect sensor cable without affecting measurement performance.

The LI200 Pyranometer sensor is designed for field measurement of global solar radiation in agricultural and meteorological studies. A silicon pyranometer puts out a current, which is dependent on the solar radiation incident on the sensor. This current output, which is directly proportional to the solar radiation, is calibrated against an Eppley precision spectral pyranometer under natural daylight conditions in units of W/m^2 and the current is measured as a voltage drop across a fixed resistor. Under most conditions of natural daylight, the error is < 5%.

The rainfall sensor consists of a gold anodised aluminium collector funnel with a knife edge that diverts the water to a tipping bucket mechanism. The mechanism is so designed that one alternate tip of the bucket occurs for each 0.025 mm of rainfall. Each tip actuates a magnetic switch. It measures rainfall at the rate of 50 mm/hour with an accuracy of ± 1%. The spent water drains out of the bottom of the housing, so the sensor requires no attention or servicing of any kind. The aluminium housing is covered with a white baked enamel surface to withstand years of exposure to the environment.

References

Anderson, K. E. and Carter, T. A. 2007. Hot Weather Management of Poultry. *Poultry Science and Technology Guide No. 30.* North Carolina State University, USA.

Arndt, C., Hazell, P. and Robinson, S. 2000. Economic value of climate forecasts for agricultural systems in Africa. *In*: Sivakumar, M. V. K. (ed.), Climate Prediction and Agriculture. Proceedings of START/WMO International Workshop, Geneva, Switzerland, September 27-29, 1999. International START Secretariat, Washington, DC, pp. 157-180.

Attri, S. D. and Tyagi, A. 2010. Climate Profile of India. India Meteorological Department, Ministry of Earth Sciences, Govt. of India, New Delhi, 122 pp.

Bajagai, Y. S. 2011. Global climate change and its impacts on dairy cattle. *Nepalese Veterinary Journal* **30**:2-16.

Barrett, E. C. 1992. Introduction to Environmental Remote Sensing. Chapman & Hall, London.

Berger, L. 1996. Emission of spin waves by a magnetic multilayer traversed by a current. *Physical Journal B* **54**(13): 9353-9358.

Bierhuizen, J. E. 1973. The effect of temperature on planr growth, development and yield. Proceedings of Uppsala Symposium, Plant Response to Climatic Factors. UNESCO, pp. 88-98.

Bisnoi, O. P. 2010. Applied Agroclimatology. Oxford Book Co., New Delhi, 540 pp.

Bourke, P. M. A. 1968. The aims of agrometeorology. *In*: Agroclimatological Methods. Proceedings of the Reading Symposium, UNESCO, Paris, pp. 11-15.

Deacon, E. L., 1969. Physical processes near the surface of the earth. *In*: Flohn, H. (ed.), World Survey of Climatology Vol. 2: General Clkimatology. Elsevier, Amesterdam.

Donn, W. L. 1975. Meteorology, edn 4, McGraw–Hill, New York. 518 pp.

Dracup, J. A., Lee, K. S. and Paulson, E.G. Jr. 1980. On the definition of droughts. *Water Resource Research* **16** (2): 297-302.

EEA, 2001. Total E. U. CH_4 emissions. European Environment Agency Factsheets.

Ghildyal, B. P. and Tripathy, R. P. 1987. Soil Physics. Wiley Eastern Ltd., New Delhi.

Hahn, G. L. 1999. Dynamic responses of cattle to thermal heat loads. *Journal of Animal Sciences* **77** (Supplement 2): 10-20.

Hartung, E. and G. J. Monteny. 2000. Methane (CH_4) and nitrous oxide (N_2O)-emissions from animal husbandry. *Agrartechn. Forsch.*, 6: E62-E69.

Hartung, J. 2003. Contribution of animal husbandry to climatic changes. *In*: Interactions between Climate and Animal Production. *EAAP Technical Series No. 7*. Wageningen Academic Publishers, Wageningen, The Netherlands.

Hoffmann, I. 2008. *Livestock Genetic diversity and Climate Change Adaptation*. Livestock and Global Change conference proceeding. May 2008, Tunisia.

Hollinger, S. E. 2009. Meteorological Forecasting for Agricultural Production. *In*: Ting, K. C., Fleisher, D. H. and Rodriguez, L. F. (eds.), Systems analysis and modelling in food and agriculture. EOLSS Publishers Co. Ltd., U. K.

Hulzebosch, J. 2006. Effective heating systems for poultry houses. *World Poultry* **22**(2): 18-19.

IMD, 2006. Trends in precipitation extremes over India. *NCC Research Report No. 3/2006*.

IPCC. 1996. Climate Change 1995: The Science of Climate Change. Contribution of Group I to the Second Assessment Report on the Intergovernmental Panel on Climate Change. Cambridge University Press, U. K.

IPCC. 2007. Climate Change 2007: The Physical Science Basis. Cambridge University Press, U. K.

IPCC. 2013. *Climate Change 2013: The Physical Science Basis. Contribution of Working Group I to the Fifth Assessment Report of the Intergovernmental Panel on Climate Change*. Cambridge University Press, Cambridge, U K.

Kiehl, J. T. and Trenberth, K. E. 1997. Earth's annual global mean energy budget. *Bulletin of the American Meteorological Society* **78**: 197-208.

Lillesand, T. M. and Kiefer, R. W. 1994. Remote Sensing and Image Interpretation, Third edition. John Wiley and Sons, New York.

MAFF. 2000. Climate change and agriculture in the United Kingdom. Ministry of Agriculture, Fisheries and Food, London.

Mavi, H. S. 1994. Introduction to Agrometeorology. Oxford & IBH, New Delhi, 281 pp.

Mavi, H. S. and Tupper, G. J. 2004. Agrometeorology: Principles and Applications of Climate Studies in Agriculture. Food products Press, Binghamton, New York, 364 pp.

Mjelde, J. W., Hill, H. S. J. and Griffiths, J. F. 1998. A review of current evidence on climate forecasts and their economic effects in agriculture. *American Journal of Agricultural Economics* **80**:1089-1095

Molga, M. 1962. Agricultural Meteorology, Part II–Outline of Agrometeorological Problems. Polish Academy of Sciences, Warsaw.

Monteny, G. J. 2003. Strategies for reducing the effects of animal husbandry on climate. *In*: Interactions between Climate and Animal Production. *EAAP Technical Series No. 7*. Wageningen Academic Publishers, Wageningen, The Netherlands.

NDMA. 2008. National Disaster Management Guidelines: Management of Flood. A publication of National Disaster Management Authority, Government of India, New Delhi, 134 pp.

NDMA. 2008. National Disaster Management Guidelines: Management of Cyclones. A publication of National Disaster Management Authority, Government of India, New Delhi, 156 pp.

NDMA. 2010. National Disaster Management Guidelines: Management of Drought. A publication of National Disaster Management Authority, Government of India, New Delhi, 78 pp.

Nix, H. A. 1981. Simplified simulation models based on specified minimum data sets: the CROPEVAL concept. *In*: Berg, A. (ed.), Application of remote sensing to agricultural production forecasting. Commission of the European Communities, Rotterdam, pp. 151-169.

Patra, A. K. 2016. Principles and Applications of Agricultural Meteorology, New India Publishing Agency, New Delhi, 424 pp.

Patra, P. K. 2012. A Text Book on Climatology. Kalyani Publishers, Ludiana, 255 pp.

Rao, G. S. L. H. V. P. 2008. Agricultural Meteorology. Prentice-Hall of India Pvt. Ltd., New Delhi. 364 pp.

Reddy, S. R. and Reddy, D. S. 2007. Agrometeorology. Kalyani Publishers, Ludiana, 313 pp.

Rosenberg, M. J. 1974. Microclimate: The Biological Environment. John Wiley and Sons, London, pp. 78-85.

Scott, M. L. and Dean, W. F. 1991. Nutrition and management of ducks. Ithaca Publishers, New York.

Sivakumar, M. V. K. 2006. Climate prediction and agriculture: current status and future challenges. *Climate Research* **33**: 3-17.

Smith, L. P. 1970. Aims and extent of agricultural meteorology. *Agricultural Meteorology* 7: 193-196.

Stigter, C. J , Sivakumar, M. V. K. and Rijks, D. A. 2000. Agrometeorology in the 21st century: Workshop summary and recommendations on needs and perspectives. *Agricultural and Forest Meteorology* 103: 209-227.

Varshneya, M. C. and Pillai, P. B. (eds.). 2012. Text Book of Agricultural Meteorology. Indian Council of Agricultural Research, New Delhi, 221 pp.

West, J. W. 2003. Effects of heat stress on production in dairy cattle. *Journal of Dairy Science* **86**: 2131-2144.

WMO. 2003. Manual on the Global Observing System, vol. I, *WMO No. 544*. World Meteorological Organisation, Geneva, Switzerland.

WMO. 2006. Drought monitoring and early warning: concepts, progress and future challenges. World Meteorological Organisation, Geneva.

WMO. 2010. Guide to Agricultural Meteorological Practices, *WMO No. 134* (Updated in 2012). World Meteorological Organisation, Geneva, Switzerland.

Yoshida, S. 1978. Tropical Climate and its Influence on Rice. *IRRI Research Paper Series 20*, IRRI, Los Banos, The Philippines.

Web Links

http://agropedia.iitk.ac.in/content/climatic-requirements
http://en.wikipedia.org
http://glossary.eea.europa.eu/terminology
http://nidm.gov.in
http://remote-sensing-agriculture.blogspot.in/agrometeorology
http://weatherwing.com/Clouds-in-Detail
http://www.agrometeorology.org/fileadmin/insam/repository
http://www.australiasevereweather.com/techniques
http://www.bom.gov.au
http://www.climatechange.gc.ca
http://www.clouds-online.com
http://www.imdagrimet.gov.in
http://www.indiaweather.gov.in
http://www.ipcc.ch/assessment□report
http://www.ipcc.ch/publications_and_data.html
http://www.moef.nic.in/sites/default/files/majumdar-incca
http://www.nasa.gov/clouds
http://www.noaa.gov
http://www.nrdc.org/health/climate/drought.asp
http://www.unisdr.org/we/inform/terminology
http://www.wmo.int/pages/themes
https://ag.tennessee.edu/solar/Earth-Energy-Budget.aspx
https://earthdata.nasa.gov/data
https://www.e-education.psu.edu

Index

A

B

C